Gerhard Pfaff
Peter Gabel
Manfred Kieser
Frank J. Maile
Joachim Weitzel

Special Effect Pigments

Technical Basics and Applications

2nd Revised Edition

Cover picture: Merck KGaA, Darmstadt, Germany

Gerhard Pfaff
Special Effect Pigments
2nd Revised Edition
Hannover: Vincentz Network, 2008
(European Coatings Tech Files)
ISBN 3-86630-905-8
ISBN 978-3-86630-905-0

Please ask for our book catalogue
Vincentz Network, Plathnerstr. 4c, 30175 Hannover, Germany
T (202) 684-6630, F (202) 380-9129
E-mail: books@american-coatings.com, www.american-coatings.com

Layout: Maxbauer & Maxbauer, Hannover, Germany
Printed by Dobler-Druck, Alfeld, HANNOPRINT, Isernhagen, Germany

ISBN 3-86630-851-5
ISBN 978-3-86630-851-0

European Coatings Tech Files

Gerhard Pfaff
Peter Gabel
Manfred Kieser
Frank J. Maile
Joachim Weitzel

Special Effect Pigments

Technical Basics and Applications

2nd Revised Edition

Preface

In recent years, effect pigments have come to play an increasingly important role in the world of industrial coloring agents. In view of this development, a reference book written by a team of authors led by R. Maisch and entitled "Special Effect Pigments" was published in 1998 by Vincentz Network.

The development of pearl luster pigments has been particularly rapid since 1998. Before then, apart from metal effect pigments, almost the only effect pigments were metal-oxide/mica pigments based on the layer-substrate principle, which according to clearly defined criteria, could be classified as "pearl luster pigments". Many of the effect pigments that have been developed and commercialized since then do not, however, easily fit the description of pearl luster pigments. This is true of materials based on Al_2O_3-, SiO_2- or borosilicate flakes as well as multilayer pigments with a Fabry-Perot structure, pigments based on liquid crystals and functional pigments with a layer-substrate structure. In recent years, the expression "special effect pigments" has become established as the generic term for all the newly developed effect pigments and an exact explanation can be found in Chapter 2.

The wide variety of uses of special effect pigments shows how important these coloring agents have become. The possibility of developing unusual colors and effects has inspired chemists and designers in the paint, plastics and cosmetics industries to give their products an individual appearance in order to enhance their marketing potential. Factors contributing to their success are not only certain special effects which are much in demand today, but also the good environmental compatibility, ease of use in application systems and outstanding durability in a pigmented medium of metal-oxide pigments and those based on Al_2O_3, SiO_2 and borosilicate.

The layer-substrate structure enables a product developer to constantly create new combinations of color and effect by exploiting not only the decorative characteristics of many new developments, but also their functional properties. Some examples are: pale, electrically conductive pigments, products with special reflection in the near IR region or pigments for the laser marking of polymers.

The importance of metal-oxide pigments is such that they are given exclusive treatment in this book. It introduces innovative pigments based on Al_2O_3-, SiO_2- and borosilicate flakes, pigments using aluminum flakes, those with a Fabry-Perot structure, and those based on polymer liquid crystals. The classic pearlescent pigments – natural fish scales, basic lead carbonate and bismuth oxychloride – are also described, as are various special types of pigment. A special chapter is devoted to

functional pigments based on the layer-substrate structure. The main properties, production methods and possible uses of all types of pigment are explained in order to show how they work. Here the experience and knowledge gained in the application technology laboratories of Merck KGaA are used to describe the possible fields of application in paints, plastics and printing inks. Color measurement is discussed in detail as it has become a very important feature of the development, quality control and practical use of special effect pigments in a wide range of systems. For completeness sake, some information on questions regarding the environment and accreditation is also provided.

Although this book is aimed in particular at color designers, application technologists and product developers in the areas of paint, plastics, printing inks and cosmetics, it is also intended for sales and marketing specialists, since it offers an introduction to the world of special effect pigments, which will be useful when preparing for meetings with customers. To sum up, this book provides a rapid but concentrated overview for those who wish to learn about this specialist area and its promising future.

Dr. Gerhard Pfaff
Darmstadt, March 2008

Content

1 History of effect pigments

Dr. Gerhard Pfaff

The luster of valuable pearls even fascinated people in the ancient world, because it differs from the normal luster of smooth surfaces with its nuances between matt and mirror luster. Pearl luster appears three-dimensional and seems to come from a depth within the object. A 3000-year-old Chinese document contains proof that humans attempted to imitate the tremendous color luster of pearls. It describes the attempts made to produce pearls with a special luster by mixing various materials [1].

Since the 11th century, experienced glass technicians of Venice produced man-made pearls with such a quality that the government felt it necessary to prohibit this. The secret of this production disappeared in the course of time with its producers [2]. First attempts to imitate pearls in France, coating small gypsum pearls with fish scale paste were unsuccessful because the body heat of the carrier tended to melt the glue fixing the scales.

The history of transparent luster pigments began 1656 in Paris when Jaquin isolated the silvery substance of the inside of fish scales and developed a suspension with silver luster [3]. This first pearl luster suspension with its brilliant luster was called "Essence d'Orient". It contained the same substance, which was later called natural fish silver. In the 17th century, buttons made of mother-of-pearl became popular and led to an upswing in the button industry in various European countries.

The development of polymethyl methacrylate and unsaturated polyester led to a new material basis for button manufacturing. Since buttons based on these polymers should further demonstrate the luster and the colored character of nacre, the search for synthetic materials in form of pearl luster pigments started and therewith the replacement of natural fish silver and mother-of-pearl, in this application and other branches. The most valuable of all newly developed pearl luster pigments proved to be basic lead carbonate. Bismuth oxychloride, which could be manufactured in form of thin platelets, grew to be the second most important synthetic pearl luster pigment.

Around 1960, a new pigment generation on the basis of metal oxide coated mica platelets was introduced to the market. The first of these pearl luster pigments to replace basic lead carbonate from its application areas were titanium dioxide mica pigments. Today there are a multitude of variations of such pigments for many decorative and technical applications on the market. The use of oxides such as iron oxides, chromium(III) oxide or mixed oxides instead of titanium dioxide permits a steady growing application of artificial pearl luster effects.

The development of new effect pigments has been continued in the last 10 to 15 years in an accelerated manner. It was decisive for this development that the possibilities for the manufacture of transparent synthetic platelets were created. These new substrate platelets are suitable for the coating with metal oxide layers in addition to mica. Platelets of silicon dioxide (silica flakes), aluminum oxide (alumina flakes), or borosilicate (borosilicate or glass flakes) belong to these new substrate materials. Additionally, synthetic mica (fluorophlogopite) is meanwhile also used instead of natural mica (muscovite) for the production of new effect pigments.

Natural mica mainly is the basis of the development of various functional effect pigments in the last years. Platelet-like electrically conductive pigments, pigments with special UV, VIS and IR properties as well as pigments suitable for the laser marking of polymers belong to this new class of materials.

New developments in the range of decorative effect pigments are multilayer pigments with Fabry-Perot structures, pigments based on liquid crystal polymers as well as pigments with structured surfaces. The history of effect pigments based on metal platelets is described in detail in the monograph "Metallic Effect Pigments" [4]. Table 1.1 contains a historical overview on the development of special effect pigments.

References

[1] *Krüger, A.*, „Perlen", Bibl. d. Unterhaltung und des Wissens, Union Deutsche Verlagsgesellschaft, Stuttgart, 1919

[2] *Simkiss, K., Wada, K.*, "Cultured Pearls – Commercialised Biomineralisation", Endeavour, **4** (1980) 32

[3] *de Keghel, M.*, „Falsche Perlen", Kunststoffe, **15** (1925) 216

[4] *Wißling, P.*, Metallic Effect Pigments, Vincentz Network Hannover, 2006

[5] *Pfund, A. H.*, "The Colors of Mother-of-Pearl", J. Franklin Inst. (Philadelphia), **183** (1917) 453

[6] *Atwood, F. C.*, US 2,278,970

[7] *Linton, H. R.*, US 3,087,828

[8] *Linton, H. R.*, US 3,087,829

[9] *Bolomey, R. A., Miller, H. A., Greenstein, L. M.*, US 3,123,490

[10] *Bolomey, R. A., Quinn, C. A., Rieger, C. J.*, US 3,437,515

[11] *Kohlschütter, H., Getrost, H., Reich, W., Rössler, H., Hörl, W.*, US 3,553,001

Table 1.1: History of special effect pigments

approx. 1655	Jaquin isolates a silvery shining substance from fish scales and begins the industrial production of pearl luster suspensions in Europe [3].
approx. 1910	Start of pearl luster pigment production in North America.
1917	A. H. Pfund discovers that the pearl luster effect is an optical phenomena based upon thin film layers lying on top of each other [5].
approx. 1920	First synthesis of pearl luster pigment suspensions containing crystalline $HgCl_2$ platelets.
approx. 1930	Various lead, arsenic and bismuth salts are synthesized for applications as pearl luster pigments.
1940	First production of platelet-shaped $PbHPO_4$ as "Iridescent Preparation P 79" by E. Merck, Darmstadt 1).
1942	F. C. Atwood describes the combination of mica and TiO_2 for the pearl luster effect [6].
1960	Basic lead carbonate ($Pb(OH)_2 \cdot 2\ PbCO_3$) achieves industrial relevance as pearl luster pigment in various applications.
1963/64	Fundamental patents for the coating of mica with metal oxides, in particular with TiO_2, using a sulfate process (thermal hydrolysis of $TiOSO_4$ in presence of mica) are granted [7–10].
1969	Cross license between DuPont, Mearl 2) and Merck regarding pearl luster pigments, including metal oxide mica pigments.
1970	Bismuth oxychloride (BiOCl) achieves industrial relevance as pearl luster pigment in various applications. The chloride process for the production of TiO_2-mica pigments ($TiOCl_2$-titration process) is developed [11].
1984	The application of metal oxide mica pigments for automotive coatings in the USA. Acceptance of this class of pigments in the entire pigment and paint industry.
1992	Effect pigments based on Al platelets coated with Fe_2O_3 are introduced to the market.
1995	Market introduction of effect pigments based on liquid crystal polymers (LCP pigments) and of multilayer pigments with Fabry-Perot structure.
1996	Introduction of effect pigments consisting of metal oxide-coated synthetic mica.
1999	Market introduction of first effect pigments based on Al_2O_3 flakes.
2000	Market introduction of first effect pigments based on SiO_2 flakes.
2002	Effect pigments based on borosilicate flakes (glass flakes) are introduced.

1) since 1995, Merck KGaA, Darmstadt/Germany
2) since 1996 affiliate to Engelhard Corp.

2 Optical principles, manufacture, properties and types of special effect pigments

Dr. Gerhard Pfaff

Pigments are substances consisting of particles which are practically insoluble in the application medium. They are used to provide color but in addition can provide corrosion inhibiting, magnetic, electrical or electromagnetic properties [1]. The range of application media for pigments includes paints, coatings, plastics, printing inks, construction materials, ceramic products, glass, enamel, and cosmetic formulations.

Pigments are differentiated and classified with respect to the way in which they interact with light. The most important pigment classes are

- white pigments,
- colored pigments,
- black pigments, and
- effect pigments.

Fundamentally, all these pigments are employed because of their inherent ability to generate color. Effect pigments are classified simply as metal effect pigments or pearl luster pigments. Effect pigments are colorants that give additional color effects, such as angular color dependence (iridescence, color travel, luster) or texture, when applied in an application medium [1]. The term "luster pigments" is also often applied to effect pigments because almost all effect pigments provide lustrous effects in practical applications. Luster pigments tend to be composed of predominantly platelet-like particles, that readily align with a parallel orientation to the surface to which they are applied. This leads to a characteristic luster arising from the reflection of incident light from the smooth surface of the pigment platelets [1].

The term "metal effect pigments" fully describes this class of effect pigments. However, "pearl luster pigments" does not give a comprehensive description of the diverse range of materials and effects provided by this class of effect pigments.

Metal effect pigments are luster pigments consisting of metallic particles only. In applications, with a parallel aligned orientation in their application system, they show a metal-like luster by reflection of light at the surface of the metal platelets. Pearl luster pigments on the other hand are luster pigments consisting of transparent or semitransparent platelets of metal oxides or other materials with high refractive index. They can also be applied with a parallel orientation to give a characteristic pearl luster generated by multiple reflections. Pearl luster pigments showing additional colors generated by interference of light are also designated as nacreous or interference pigments [1]. Interference pigments are by definition effect pigments whose color is generated completely or predominantly by the phenomenon of interference of light [1].

Various new effect pigments consisting of platelets of transparent materials have been developed in recent years. These pigments exhibit effects arising from multiple reflection followed by light interaction, but on the other hand they do not have a characteristic pearl luster. In addition, the class of effect pigments has been broadened by the development of interference pigments based on platelets of non-transparent materials. This has made the classification more complex and therefore the new term "special effect pigments" was created. This includes all platelet-like effect pigments, which can not be categorized as metal effect pigments [2]. According to this categorization, special effect pigments are pearl luster pigments and transparent and non-transparent interference pigments, which independent of their composition and structure lead to pearl lustrous or non pearl lustrous effects in combination with interference phenomena in the application medium.

Special effect pigments are natural or synthetic pigments, characterized by high luster, brilliance and iridescent color phenomena from optically thin films. The visual impression has its origin in reflection and refraction of the light at thin single and multiple layers. It is not limited in nature to pearls and clamshells as it can be seen in many fascinating examples in the world of birds, fishes, gemstones, minerals and insects. Fundamental investigations into the optical principles of the luster of natural pearls for instance shows that the brilliant colors are derived from layered and structured biopolymers formed by bio-mineralization [3–5].

Figure 2.1 shows a comparison of the optical principles for the interaction of light with white, colored and black pigments, metal effect pigments and transparent interference pearl luster pigments.

White, colored and black pigments typically have particle diameters of 0.1 to 1 µm. This encompasses the wavelength range of visible light. Somc carbon black pigments have diameters even below 0.1 µm. Effect pigments on the other hand have particle sizes predominantly in the range 5 to 100 µm, in some cases also above these values. Therefore they have an average platelet diameter significantly larger than the wavelength of visible light. The thickness of the platelets, with some minor exceptions, is typically below 1 µm, which is of the order to interfere with visible light. Effect pigments therefore have high aspect ratios (ratio of diameter to thickness), with values of up to 200.

White, colored and black pigments are mostly irregularly formed and predominantly interact with light as follows (Figure 2.1).

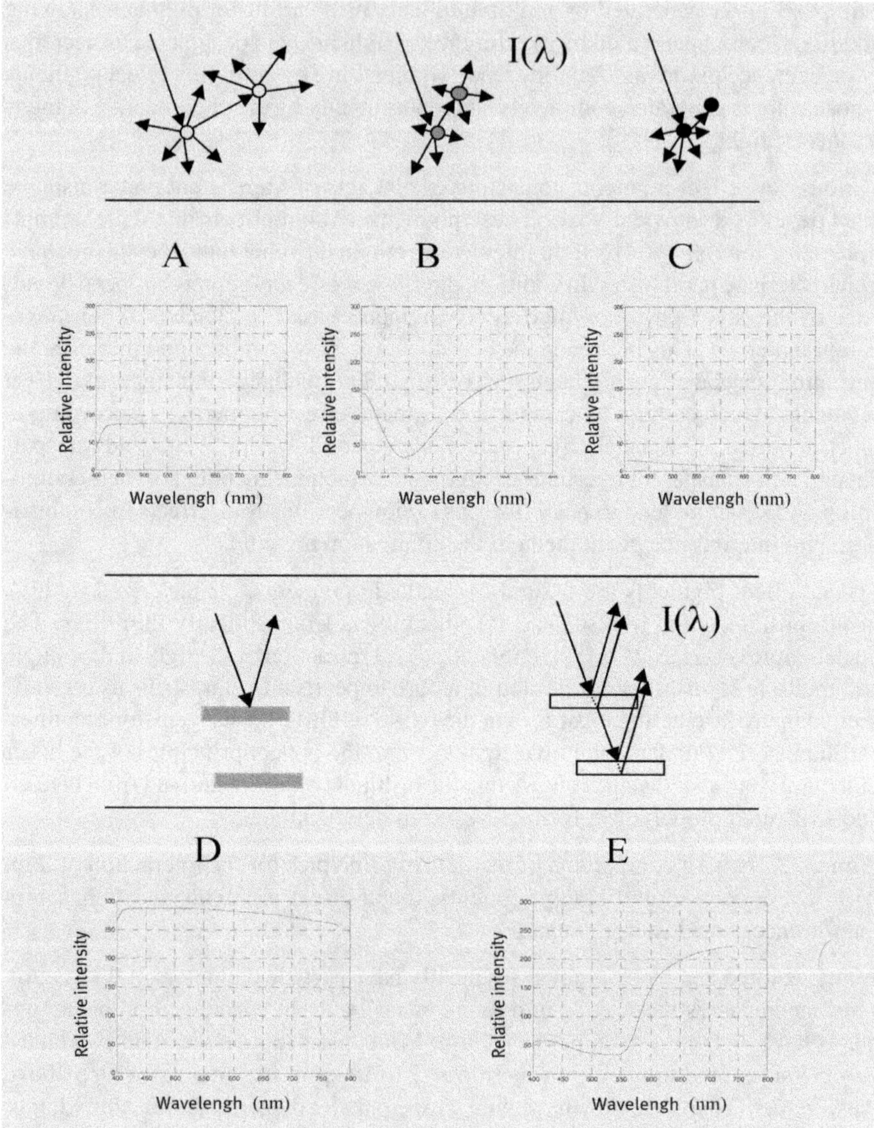

Figure 2.1: Schematic description of single particles of the types: (A) white pigment, (B) colored pigment, (C) black pigment, (D) metal effect pigment, and (E) pearl luster pigment in an application medium (e.g. paints, plastics, printing inks, cosmetic formulations). The typical interactions are shown, which lead in a large ensemble of particles to the below shown exemplary remission spectra (a–e). They are also the basis for classifying the pigments. (a) titanium dioxide (white), (b) chromophthal red, (c) carbon black, (d) aluminum (metal effect), (e) rutile on mica (interference red).

White pigments (A): The pigment particles completely scatter the incident light equally in all directions (diffuse reflection).

Colored pigments (B): The pigment particles absorb certain wavelengths of visible light and scatter the remaining wavelengths, depending on the color (selective absorption).

Black pigments (C): The pigment particles absorb all wavelengths of visible light (complete absorption).

Metal effect pigments (D) consist of thin metal (mostly aluminum) platelets; they are not transparent and reflect all incident light in one direction. The pigment particles act like small mirrors leading to a reflecting metal luster (metallic effect) when applied to give an aligned parallel orientation. Transparent interference **pearl luster pigments** (E) selectively reflect part of the light that is directly incident on the smooth platelet surface. The other part of the light enters into the transparent particles and is partially reflected either at interfaces within the pigments or at the bottom of the platelets. The internally reflected light leaves the platelets leading to interactions with adjoining pigment particles to create further reflections (multiple reflection). The partial reflection of light and optical superimposition create interference phenomena, which together with multiple internal reflections lead to the characteristic appearance of pearl luster pigments. The difference in refractive index between the high-refractive index pigment (typical values 1.8 to 2.9) and the application medium (the latter one has mainly values in the range from 1.5 to 1.6) is essential to the optical processes involved.

Pearl luster pigments can imitate the luster of natural pearls when the pigment layer and the paint layer structure are chosen in a suitable manner. Pearls and mother-of-pearl consist of a multitude of alternating transparent layers with different refractive indices. They have in most cases much larger lateral dimensions than in pearl luster pigments. The layers consist of calcium carbonate ($CaCO_3$) with a relatively high refractive index and proteins with a relatively low refractive index. The optical effects seen in pearls, mother-of-pearl and pearl luster pigments are similarly created by the interference of light rays, which are at least partly reflected at different interfaces.

Differences in the refractive indices of adjoining thin layers, particularly between pigment layers and the surrounding medium are a prerequisite for all iridescent colors arising from interference effects. Such effects are also well-known from the interfaces between air and an oil film or an oil film and a water surface. Similar differences in refractive index arise in the use of special effect pigments where the high refractive index platelet layers border upon the low refractive optically thinner application systems such as lacquers, printing inks or plastics. In applications, with a parallel aligned orientation of the platelets in the medium, by selective choice of the layer thicknesses of the pigments the interference color and effects can be controlled in a manner suited to the wavelength of the incident light.

The basic types of pearl luster pigments are synthetic and natural pearl luster, interference pigments, transparent and partly absorbing. The pigments can consist of one layer (e.g. BiOCl) or of several layers with different refractive index (e.g. TiO_2 on mica). They can be made up of platelets with the same or varying thickness. In almost all cases, they are composed of crystalline layers (e.g. $Pb(OH)_2 \cdot 2\,PbCO_3$, BiOCl, TiO_2 on mica).

2.1 Optical principles

2.1.1 Reflection and refraction at one and at several interfaces

The human eye can perceive objects only when electromagnetic waves with wavelengths in the range of 380 to 780 nm (visible light), which contain information about the objects execute analyzable color appeal. This is only possible when the object is emitting light by itself (self-illuminator) or when it is radiated from a primary light source and the incident light is thereby modified in regard of direction, intensity and spectral composition or of its phase (non-self-illuminator). The changes in the direction of light rays when striking an object with dimensions ($l >> \lambda$) are described by the law of reflection and Snell's law of refraction, whereby the refractive index of the medium that the light ray exits and that of the object that is entered must be different. According to Figure 2.2, the following optical laws have to be considered for the description of the light path:

- the law of reflection, where the angle of incidence ε and the angle of reflection ε' are identical

(1) $\varepsilon = \varepsilon'$

- and Snell's law of refraction

(2) $$\frac{\sin(\varepsilon')}{\sin(\varepsilon'')} = \frac{n_1}{n_2}$$

The refractive indices n_1 and n_2 represent a relative measure of the velocity of light v in the related medium. The refractive index of a vacuum is defined as $n_v = c/v = 1$. Therefore the refractive indices of all other materials have values greater than 1, because the related light velocities always stay below the value in vacuum. Some examples of refractive indices of different materials are summarized in Table 2.1. The existing dependency of the refractive index from the wavelength of the light and from temperature can be neglected for these considerations. For a first approximation, it is possible to rely on averages or to use the values, which are effective for the perpendicular direction.

According to the law of refraction (2), the non-reflected part of the incident light (neglecting absorption and scattering) is deflected during the entrance in the optically denser medium in the perpendicular direction ($\varepsilon'' < \varepsilon'$), see Figure 2.2. When light, on the other hand, is entering an optically thinner matter, the angle is enlarged in the perpendicular direction ($\varepsilon'' > \varepsilon'$).

Incident P1 - P0, reflected P0 - P2 and refracted P0 - P3 light rays are always located in the same plane. Both laws are the result of the Fermat principle, which explains that the paths of the light rays between P1 and P2 respectively P1 and P3 are configured in such a manner that the light needs the least time to travel this distance.

Laws (1) and (2) are valid also in cases when the incident light rays arrive from different directions, the interface itself is formed irregularly and consists of non-ordered single elements ($d > \lambda$) or several boundaries in succession are not ordered in a plane-parallel manner. All of these cases can often occur separately or in combination (e.g. during diffuse illumination of paint surfaces, at paint surfaces with considerable unevenness and at paint layers with disoriented, preferably transparent and semitransparent interference

Table 2.1: Refractive indices of various materials [6]

Medium	Refractive index
vacuum, air	1.0
water	1.33
MgF_2	1.37
SiO_2	1.42 to 1.48
plastics, paints, printing inks	1.4 to 1.7
mica	1.5 to 1.6
Al_2O_3	1.7
guanine, hypoxanthine	1.85
basic lead carbonate	2.0
bismuth oxychloride	2.15
Fe_3O_4 (magnetite)	2.4
TiO_2 (anatase)	2.5
TiO_2 (rutile)	2.7
α-Fe_2O_3 (hematite)	2.9

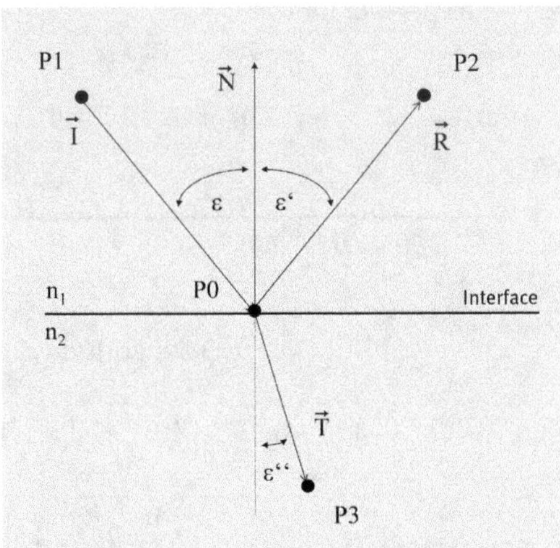

Figure 2.2: Schematic description of the direction of the rays after light reflection and light refraction at the interface, which divides two media with the refractive indices n_1 and n_2. $\vec{I}$, $\vec{R}$ and $\vec{T}$ are the direction vectors of light extension in the incident, the reflected and the refracted ray. N is the perpendicular vector of the interface.

pigments) and lead to a "diffuse reflection". However, the light rays are now not more necessarily located in one plane and the directions of propagation of the light rays are described as vectors in space (see designation in Figure 2.2) with the corresponding law of reflection

(3) $\vec{R} = 2\vec{N}\cos(\varepsilon) + \vec{I}$

and with the law of refraction

(4) $\vec{T} = \left(\frac{n_1}{n_2}\right)\vec{I} - \left(\cos(\varepsilon'') - \left(\frac{n_1}{n_2}\right)\cos(\varepsilon')\right)\vec{N}$

(see Figure 2.2 as well).

When the illumination of the object is by many light rays at the same time, the limiting cases, which have to be distinguished are either the completely non-directed (diffuse) or the completely directed incident radiation. Diffuse illumination reflected at a plane interface remains diffuse. Directed illumination, on the other hand, is reflected diffuse when the interface is formed irregularly, is consisting of non-ordered single elements or when the light is deflected by scattering processes.

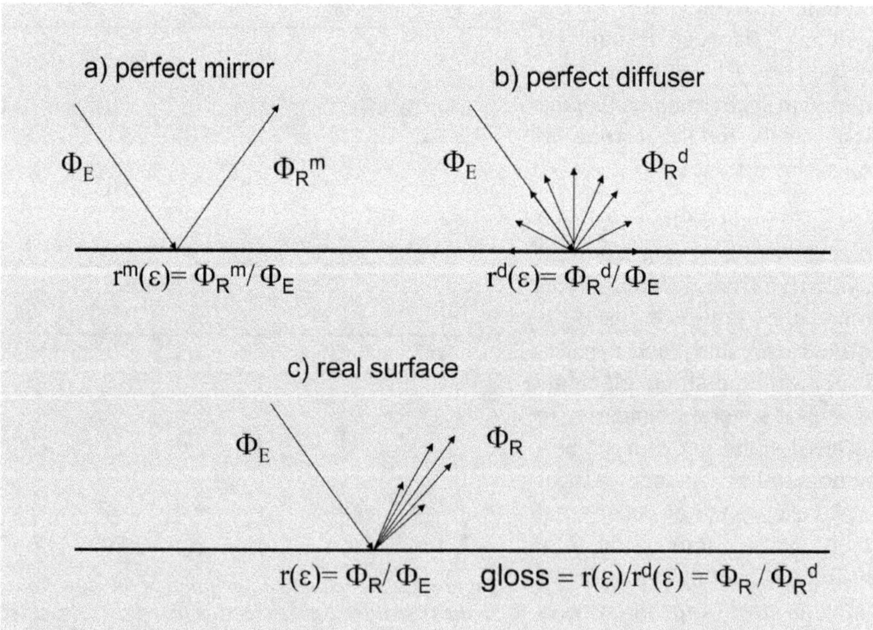

Figure 2.3: Schematic description of the borderline situations for the reflection respectively the scattering of light: (a) perfect mirror (mirror reflection), (b) perfect diffuser (only light scattering), (c) real surface (the reflection indicatrix exhibits a geometric form defined by scattering and reflection processes).

The comparison of different materials with different interface structures shows that rough, uneven surfaces reflect light rays diffusely in all directions. On the other hand, even surfaces reflect in the ideal case all light rays in one preferred direction. There are in practice, e.g. in paint systems, flowing transitions between situations a) (perfect mirror, high reflection coefficient) and b) (perfect diffuser, lowered reflection coefficient compared with a) described in Figure 2.3. Shape and size of the reflection indicatrix, in other words of the spatial spectral intensity distribution of the reflected radiation, include now all relevant reflection characteristics of the sample, which have their origin in scattering and reflection at the various boundaries – starting from the clear varnish to the pigment particles located deep in the system that can still interact with incident light.

The reflection coefficient r(ε) is now defined as the ratio of the reflected light intensity I_R to the incident light intensity I_E while the gloss G is determined as the ratio r_g (sample)/r_g (perfect diffuser) in the specular direction.

High gloss values are therefore also the reason for the strong and angle-dependent brightness, which not only appears at polished metal surfaces but also at boundaries of transparent bodies.

The reflection of directed light on even non-metallic surfaces (pure regular reflection) can be formulated mathematically using Fresnel`s reflection coefficient r(ε). Its description for non-polarized light is expressed by the following equation:

$$(5)\ r(\varepsilon) = \frac{I_R}{I_E} = \frac{1}{2} \cdot \left[\frac{tg^2(\varepsilon' - \varepsilon'')}{tg^2(\varepsilon' + \varepsilon'')} + \frac{\sin^2(\varepsilon' - \varepsilon'')}{\sin^2(\varepsilon' + \varepsilon'')} \right]$$

Figure 2.4 illustrates the relationship between the reflection coefficient r(ε) and the angle of incidence ε for non-polarized light at $n_2 = 1.5$ and $n_1 = 1$. For angles of incidence below 30°, the values for r(ε) are almost constant at 0.04. For this reason, in this range the simple equation for light, arriving from the perpendicular direction (ε = 0°), can be used

$$(6)\ r(0°) = \left[\frac{n_2 - n_1}{n_2 + n_1} \right]^2$$

That means that a luminous flux coming from air at almost vertical contact with water will be reflected by 2%, with lacquer by 4%, with basic lead carbonate by 11 %, and with rutile by 21%. Figure 2.5 just shows this dependency r(ε = 0°) as a function of n_2 at $n_1 = 1$.

Only the difference in refractive indices is important for the size of the reflection coefficient, regardless whether the luminous flux comes from a more or less dense medium. If the second medium has a higher refractive index, the sign of the electrical field vector changes for the reflected part of the light, i.e. the light is shifted by

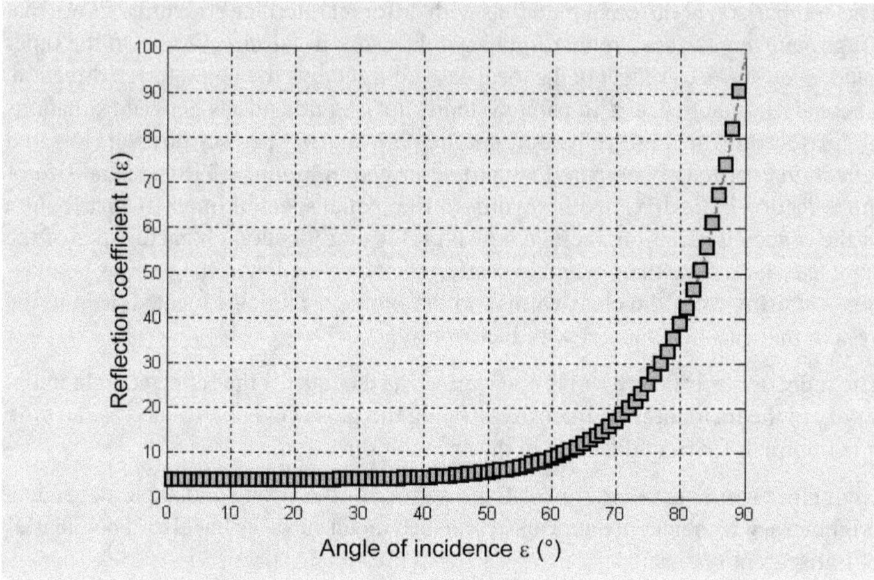

Figure 2.4: Reflection coefficient r(ε) as function of the angle of incidence ε at $n_2 = 1.5$ and $n_1 = 1.0$ (according to equation (5))

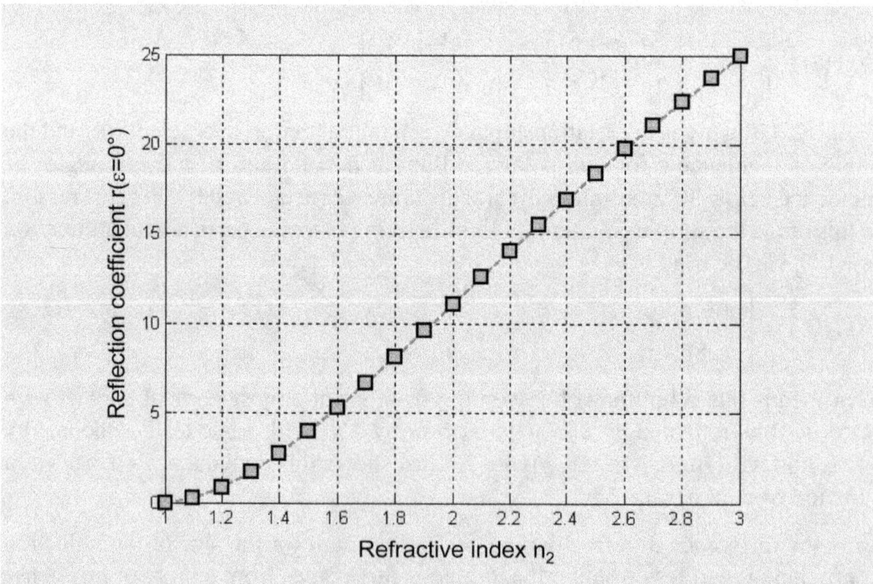

Figure 2.5: Reflection coefficient r(ε = 0°) as function of the refractive index n_2 at $n_1 = 1.0$ (according to equation (6))

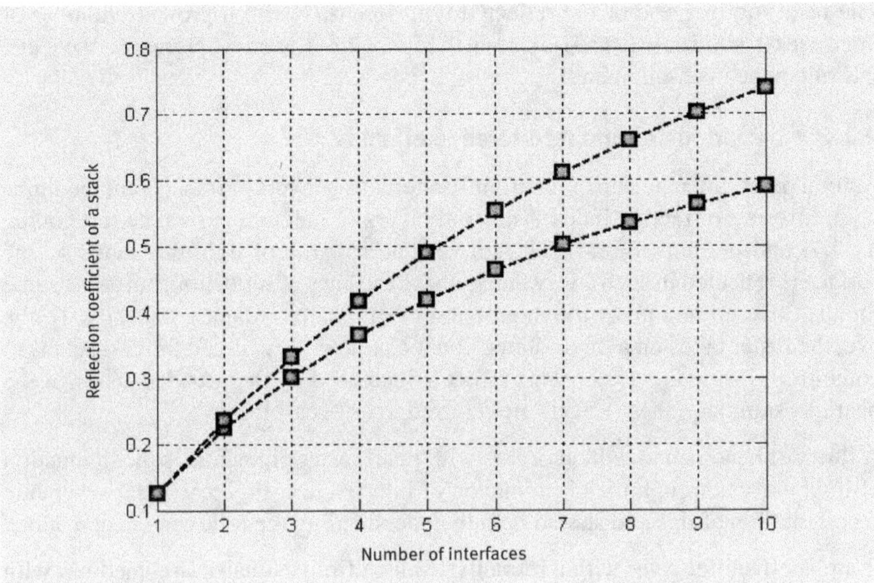

Figure 2.6: Reflection coefficient r(ε = 0°) with a growing number of interfaces in a stack. The upper curve is calculated according to equation (7), the lower curve according to equation (8). The curves correspond calculations without and with multiple reflections at a stack of k optically not coupled interfaces.

half a wavelength in phase length. At interfaces to substances with a lower refractive index no phase shift takes place. This phase shift for reflection at optically denser media plays a significant role also for the interference behavior of layers.

Generally speaking, only a small part of the light meeting the interface will be reflected, the majority enters the next medium or passes through it. If the initial luminous flux I_E is 1, then the reflected flux I_R is given by the reflection coefficient r. If the light continues through a fully transparent and clear medium and then meets another interface, the principle occurring on the first interface repeats itself. The light intensity striking the second interface is now of course already diminished by the reflection at the first interface. Therefore the reflection coefficients of the interface order k, r(k) relates to that in the first interface as

$$(7)\ r(k) = 1-(1-r)^k$$

The following equation describes the situation considering multiple reflections between k parallel interfaces:

$$(8)\ r(k) = \frac{kr}{1+(k-1)r}$$

The resulting increase of the reflected light intensity with a growing number of interfaces (for the case ($\varepsilon = 0$) is shown in Figure 2.6. For effect pigments, the value k is of course even-numbered.

2.1.2 Pearl luster and interference effects

Natural pearls are made up of recurrently alternating layers of transparent mediums with differing refractive index. They are layers of calcium carbonate (aragonite, $n \approx 1.7$) and protein ($n \approx 1.4$). Thus, a very narrow ray of light that lands on the surface is reflected in such a way that it obeys the laws of reflection, but at the same time is diffused and gives the impression of gradually reducing intensity. To the eye, the luster takes on a three-dimensional character, because at times it seems to come from inside the object. This effect is increased by the rounded shape of the pearl. In summary, these effects are referred to as "pearl luster".

A further phenomenon, which occurs with pearl luster pigments, is the interaction of light waves through the phenomenon of interference. Figure 2.7 is a schematic representation of this and should help in understanding the following explanation.

If a plane-parallel layer with a refractive index of n_2 is situated in a medium with a different density, then $n_1 = n_3$. It has already been mentioned that a light ray (L),

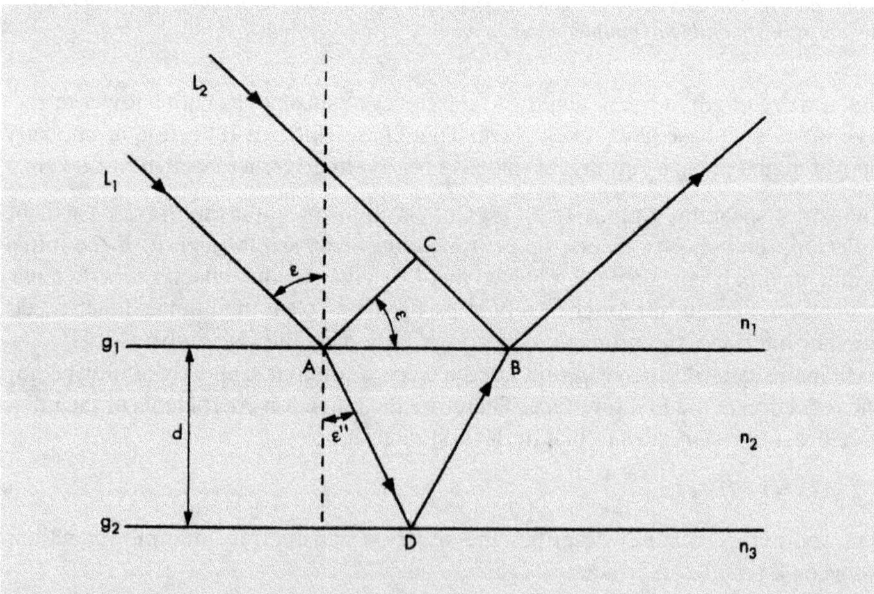

Figure 2.7: Schematic diagram showing the phase relationships of interfering light rays (L_1 and L_2). The "detour" for L is AD + DB – CB.

which is reflected at the interface with an optically denser medium (g_1) experiences a phase shift of half a wavelength or 180°. However, at the interface with a material with a lower refractive index (g_2), no phase shift takes place. The $\lambda/2$ phase shift in the first example is shown in Figure 2.8 as a sinusoidal oscillation.

The two reflected separate rays L_1 and L_2 interact with one another when superposition occurs. This leads to visible interference effects. A path difference of the rays occurs owing to the phase shift $\lambda/2$ of L_2 and the differences in geometry and time of the path of L_1. The diverted path of the ray L_1 is $2d \cdot n_2$ when light falls vertically. The product $d \cdot n_2$ (generally referred to as $d \cdot n$) is called the "optical path length".

How interference happens

Considering sinusoidal oscillation of two rays, a path difference can lead to a situation that a peak of wave in one of the divided rays meets the wave trough of the other one (Figure 2.9). This situation occurs under the following conditions:

(9) $2d \cdot n_2 = \lambda \cdot x$

where $x = 1, 2, 3, ...$; λ = wavelength of the divided ray.

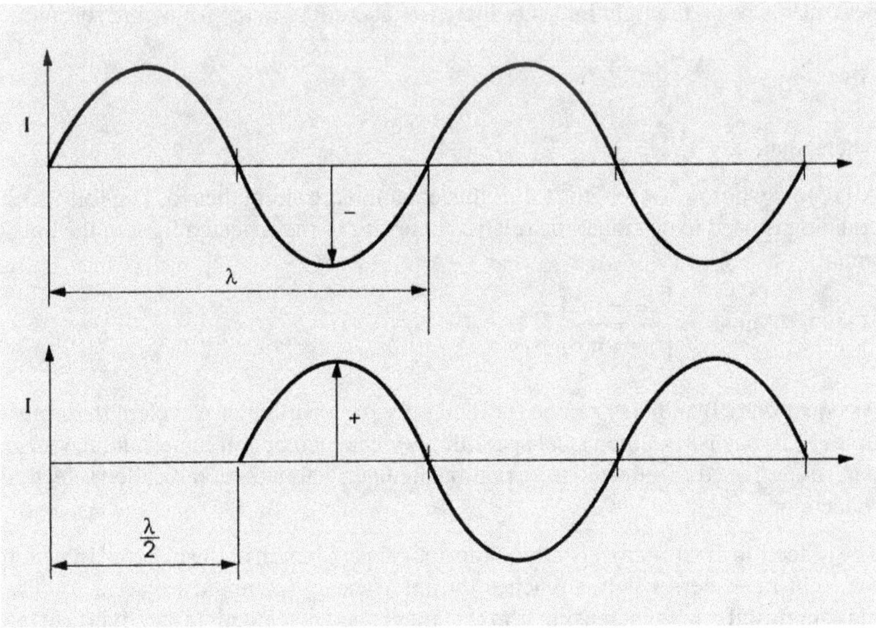

Figure 2.8: Phase shift of a half wave length by a sinus wave

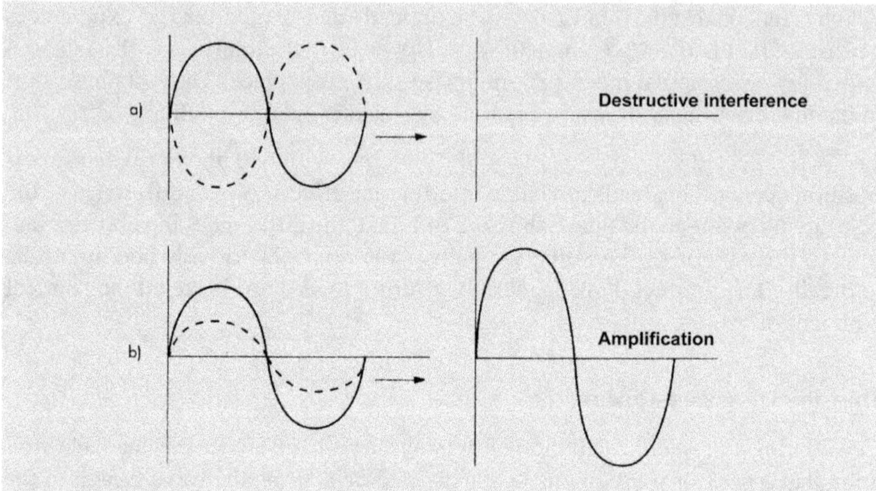

Figure 2.9: Interaction between light rays (pictured as sinus waves) with various phase lengths: a) phase crest to phase trough, b) phase crest to phase crest

In this situation, the light intensity reduces. All the colors whose wavelengths fulfill this equation are eliminated from the spectrum of white light. On the other hand, if peak meets peak, the light intensity increases according to the following formula:

$$(10) \quad 2d \cdot n_2 = \frac{\lambda(2x+1)}{2}$$

where again x = 1, 2, 3, ...

All colors whose wavelengths fulfill this equation are strengthened. The following equation is used to calculate the relative intensity of the reflected light in the maximum:

$$(11) \quad r(0°)max = \left[\frac{n_2^2 - n_1^2}{n_2^2 + n_1^2}\right]^2$$

As equations (9) and (10) can be fulfilled only for a particular wavelength or multiples of it, one part of the visual spectrum is weakened or eliminated and the other part strengthened or reflected in an optimal manner. This situation is called selective reflection.

The reflection maximum depends on the optical path length of the material in which the light ray enters when it reaches the interface g_1 (compare Figure 2.7). The maximum shifts with increasing film thickness and wavelength respectively. In the case of a layer with an optically high refractive index, which has material either

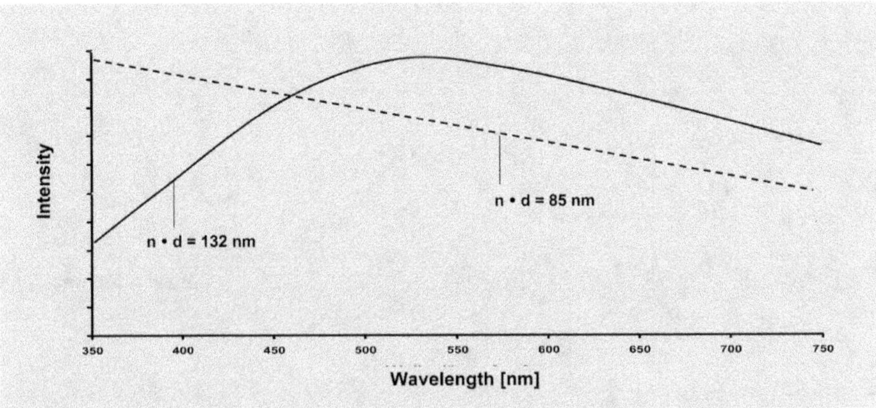

Figure 2.10: Relative distribution of intensity during reflection of various thick oil films

side of it with a lower refractive index ($n_1 < n_2 > n_3$), then the reflection peak n · d is under 85 nm in the ultraviolet range. The highest luster or almost white reflection is observed at optical wavelengths between 85 and 140 nm. It is, however, necessary to take into consideration the fact that this part of the visible spectrum is also weakened by interference. A clear example of this can be seen in the curve of relative spectral intensity distribution of light that is reflected from thin layers with optical path lengths between 85 and 132 nm [7], Figure 2.10.

The result is that the thin layers can never reflect purely white light. However, in the area of maximum luster mentioned above, a blue or yellow nuance is visible only under close scrutiny, because of the flatness of the curve.

As the layer thickness increases, the reflection maximum in the visible part of the spectrum is accompanied by a minimum and noticeable color differences occur in the order yellow, red, blue and green. If the optical path length is increased with 85 nm as a starting point, the reflection maximum first enters the yellow and red spectral areas. As the increase of the product n · d continues, the sequence of the colors repeats itself in the second order. Beginning with blue, the second order colors green, yellow and red are observed. In a similar way, colors of even higher orders occur as the optical path length continues to increase. As the layer thickness grows, the maxima and minima follow one another increasingly closely. The result is also that the colors in the different orders vary in shade and intensity.

Figure 2.11 shows the calculated intensities, at optical path lengths n · d of 322 and 538 nm, of reflected rays of light at various wavelengths. In both curves, the maximum occurs at approx. 430 nm, i.e. in the blue area. When the layer is thin, the interference color is pure blue, but with thicker films, the presence of the second reflection peak in the visible range at about 710 nm gives it a noticeable reddish hue.

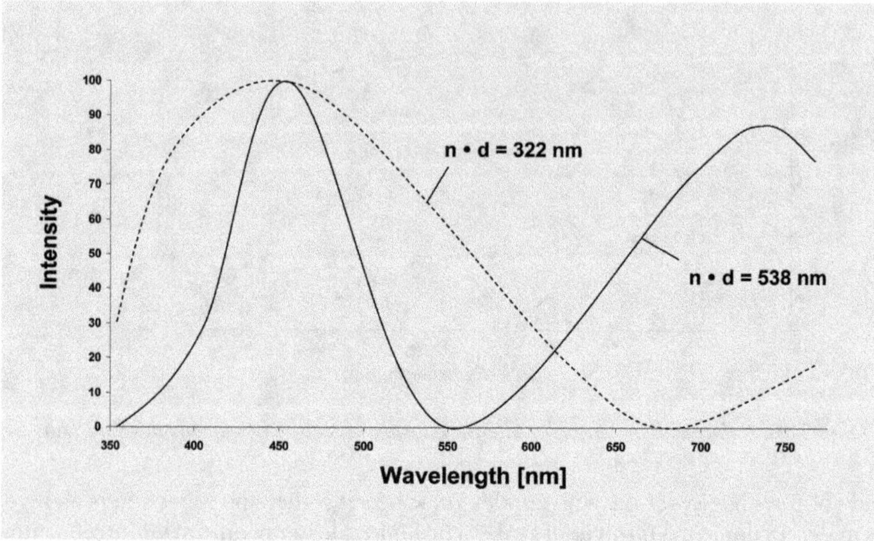

Figure 2.11: Relative distribution of intensity during reflection at layers that correspond to blue of 1st order and blue of 2nd order

During the manufacture of interference pigments, this phenomenon can be illustrated in different way by using so-called deposition curves (these curves show the successive growth of the thickness of an interference layer starting from d = 0 by the color development). Figure 2.12 shows the curve for a material with a refractive index of n = 2.15. The curve begins in the achromatic area of the x,y-color diagram at $n \cdot d = 0$. As the optical path length increases, the curve develops clockwise from the starting point in a spiral form round the achromatic point. The curve moves through color orders 1 and 2 and higher orders. Depending on the color location, either the first or second order can be saturated. Coordinate Y (luminance) is set at zero here to make things clearer. For the purpose of studying the curve, the refractive index of the medium is assumed to be 1.45. This is a typical value for a binder.

Thin layer interference is the splitting of light. Thus the transmitted light is always complementary to the reflected light. In addition to these complementary colors, a further feature of thin-layer interference is the dependence of the color on the angle at which the light strikes the object. When this angle is varied from 0 to 90°, the color coordinates in the representation as x,y tend to result in an anti-clockwise shift.

Mixtures of interference pigments (in general, interference layers) behave according to the additive color mixing principle (in contrast to absorption pigments which behave according to the subtractive color mixing principle). This means that in coor-

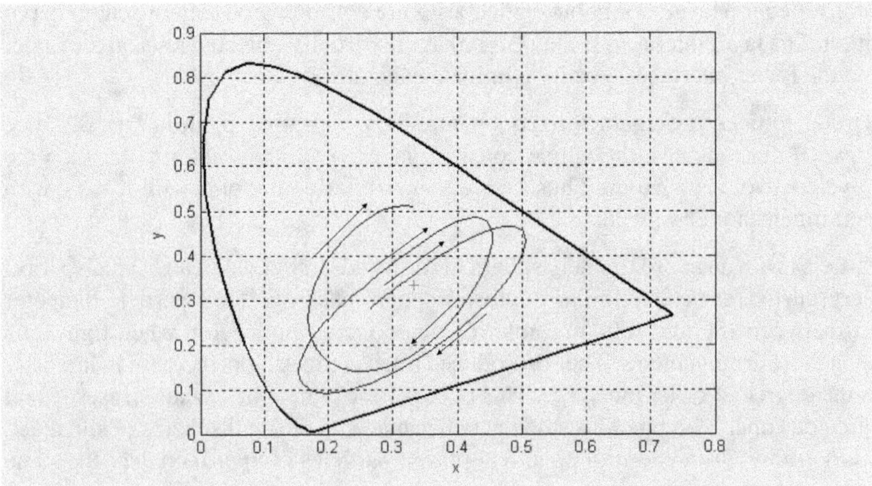

Figure 2.12: Typical curve for the development of the color coordinates x and y of an interference layer with increasing optical path length (n_2 = 2.15, the direction of the thickness increase is marked by arrows). The periodicity for achieving determined hue angles, although with different chroma, is evident.

dinates x, y and Y used in Figure 2.12, the values of mixtures always lie on the straight lines connecting the x, y and Y initial interferences of the two mixed colors.

Occurrence of interference colors

Interference colors, or iridescent colors, Newtonian colors or colors of thin films as they are also known, often occur both in nature and the world of science. Soap bubbles and thin films of oil are well-known examples of the incidence of interference colors. Interference effects are best seen with a dark background, because it absorbs the wavelengths of the color, which is complementary to the reflected color and also much of the light, which lands on the object. In contrast, a pale background reflects the light and the effects are no longer visible.

Iridescent costume jewelry is popular. Jewelry manufacturers use a number of techniques to make thin films for interference effects, so they do not have to depend on minerals to give iridescence. The oldest method is the application of a very thin film of nitrocellulose. This is achieved by dipping an object in a dilute solution of cellulose nitrate. However, this alone is not sufficient to produce the required effect. An improvement can be made by applying alternate layers of material with significantly different refractive indices. It is relatively easy to produce a layer with high refractive index by using the hydrolysis of titanium alkoxides. However, reproducibility is a problem. An interesting technique is the high-vacuum deposition of certain materials such as magnesium fluoride, silicon monoxide or zinc sulphide [7].

Anti-reflection layers for photo-optical uses are also prepared using vacuum deposition. The layer thickness is chosen so that the partially reflected rays on the upper and the lower interfaces equally eliminate each other.

Crystal lattices are characterized by regularly repeating planes. The distance between these planes, the lattice constant, is almost 1000 times smaller than the wavelength of visible light. Thus, crystals show interference only with X-rays, with their much shorter wavelengths.

Particles with an approximately spherical form and similar diameters can also form a crystal-like structure similar to closest sphere packing. If the particle diameter lies between 150 and 1000 nm, interference effects can develop when light falls upon these arrangements. The so-produced interference colors increase in intensity as the degree of order increases. This occurs for example in certain viruses [8] and "precious opal" [9], a precious stone which is made up of small spheres of silica gel. Furthermore, dispersions of spherical plastic particles of approximately the same size and arrangement and with the appropriate diameter show opalescence [10].

The dimensions of cholesteric liquid crystal structures lie in the area of the wavelengths of visible light and thus produce interesting color effects. In the liquid crystals, the individual molecules arrange themselves with their longitudinal axes at a particular angle, therefore forming themselves into a screw-like arrangement. The distance between those molecules whose longitudinal axes are parallel to one another is several hundred nanometers. The lattice distances vary with temperature which, in turn, causes a color change within the cholesteric liquid crystals [11, 12].

2.2 Natural pearl luster pigments

The first pearl luster pigment was the so-called "natural fish silver". As the name infers, it was derived from the scales of fish [13–17]. The most prolific sources of fish scales are: herring, carp, and sardine. They constitute less than 1% of natural guanine and their extraction is difficult. Hence, their cost is high. In the dry state, the pigment tends to agglomerate and is also very brittle. For this reason, the pigment is supplied only as a 22–25 % solids suspension under the name Pearl Essence or Essence d'Orient.

Natural fish silver consists of the two purines: guanine (75–97 %) and hypoxanthine (3–25 %).

The substance crystallizes as needles or as long platelets that are 1–10 µm wide, 20–50µm long, and have a thickness of only 0.025–0.075 µm (see Figure 2.13). The optically anisotropic crystals have a refractive index of 1.8 and a density of 1.6 g/cm^3. The latter property of a relatively low density is the reason why guanine, despite its expense, is still used today in cosmetic formulations. It is especially used in nail varnishes as the pigment particles hardly settle out at all.

Fish silver pigments are by-products of the fishing industry. The pigment crystals are extracted from an aqueous suspension of the fish scales using organic solvents. During this process, the proteins are dissolved and the crystals become separated from the fish scales. Final isolation of the guanine and/or hypoxanthine crystals is done by centrifuging, followed by washing to remove any last traces of fish oils or soluble proteins [15].

The ratio of guanine and hypoxanthine varies with the type of fish and the geographic origin (usually Japan, Norway, northeast coast of the USA, Canadian coast). Guanine and other nitrogen containing heterocyclic compounds can also be produced synthetically in lamellar form [18–20]. Until the present date no synthetic products have appeared on the market.

An interesting effect can also be achieved with "pearl flakes" prepared from ground mussel shells. For example, they are made by grinding the waste material from mother-of-pearl button production. The particles obtained in this way are so coarse that they can hardly be called a pigment.

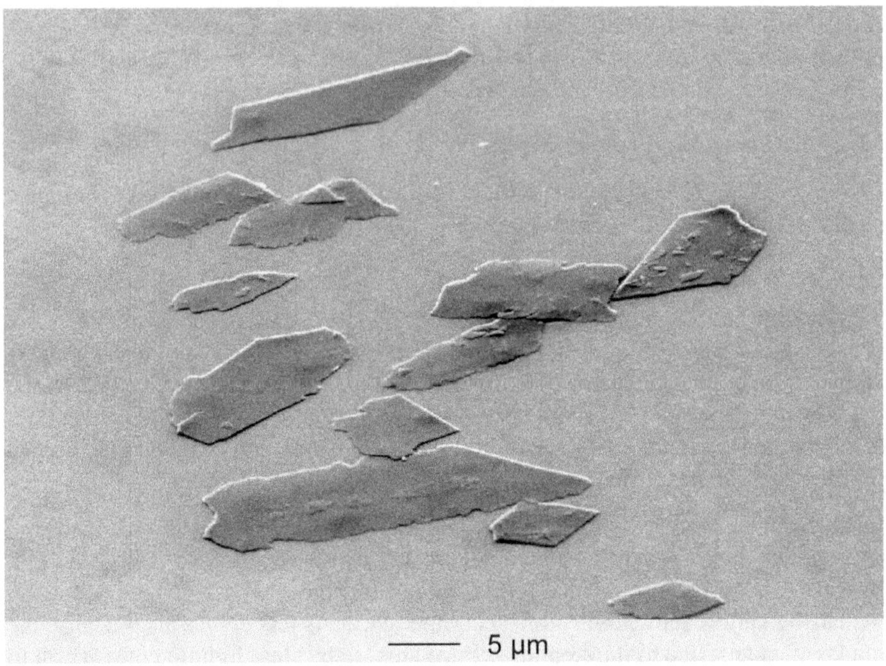

Figure 2.13: Scanning electron micrograph of natural fish silver

2.3 Basic lead carbonate

The physical properties of synthetic pearl luster pigments, such as basic lead carbonate, have to meet the following requirements:

a) *High refractive index (to achieve high reflectivity)*
The main application media for pearl luster pigments are paints, plastics, and cosmetic formulations with refractive indices between 1.5 and 1.6. The refractive index should lie above 1.8 to achieve a distinct effect. Many inorganic compounds meet these requirements and considerably surpass the quality of organic natural fish silver.

b) *Good transparency without self-color (to achieve pure interference colors)*
This requirement notably narrows the selection of suitable materials.

c) *Platelet structure with smooth surface (for optimal directed reflection)*
Most substances do not crystallize spontaneously in the form of smooth platelets. In certain cases, the crystallization can be influenced to build up platelets.

d) *Optimal thickness*
Silver white pearl luster pigments reach their maximum reflection only when the optical thickness of the particles lies between 85 and 140 nm.

A pearl luster pigment that ideally matches all four requirements is basic lead carbonate, manufactured using special methods. The general formula is $Pb(OH_2) \cdot 2\ PbCO_3$. It crystallizes in well-shaped hexagonal platelets [18, 21].

Manufacturing and properties

$Pb(OH_2) \cdot 2\ PbCO_3$ is precipitated from aqueous lead acetate or lead propionate solutions by blowing in carbon dioxide:

$$3\ Pb(OAc)_2 + 2\ CO_2 + 4\ H_2O \rightarrow Pb(OH)_2 \cdot 2\ PbCO_3 + 6\ HOAc$$

Temperature and pH value must be controlled precisely. During an optimal reaction process, hexagonal platelets are formed (approx. 50 nm thick, diameter approx. 20 µm). The high refractive index of 2.0, a high aspect ratio of more than 200 and the extremely smooth surface give basic lead carbonate an incomparable luster. As demonstrated in Figure 2.14, preparations with these pigments consist of hexagonal platelets with an exact shape, constant size and very smooth surface.

If the thickness lies at 40 nm, a blue metal like luster develops, at 70 nm, a yellowish, warmer shine appears. Even thinner crystals do not reflect enough visible light, furthermore the platelets are very brittle and sensitive. Thicker platelets show an increasing yellow reflection color. Their application is less economic, because the thicker platelets, due to the smaller surface area, reflect less light in comparison to the quantity of pigment.

There is also an optimal range for the platelet diameter. Crystals, which are too small, exhibit a high number of edges, scattering the incident light. Small crystals also limit the parallel alignment in the application medium necessary for an optimal optical effect. Larger particles, on the other hand, demonstrate a poorer mechanical stability. If larger than 25 to 30 μm, they can give a visibly grainy surface structure in the pigmented layer.

The high density (6.4 g/cm^3) of basic lead carbonate is also disadvantageous, leading to quick sedimentation in the application medium. Further disadvantages are the relatively limited chemical and thermal resistance, which prohibits an application with acids or sulfides, as well as heating to temperatures above 200 °C.

Basic lead carbonate is still used frequently for the manufacture of buttons and artificial pearls. However, it is very controversial due to its lead content, and Merck KGaA, amongst other companies, has completely ceased its production. The final decision regarding the future of this pigment will be made by the consumer and the regulatory authorities concerned.

Other lead-containing pigments

In the history of the development of synthetic pearl luster pigments, other lead compounds have played a role, including lead hydrogenphosphate $PbHPO_4$, lead

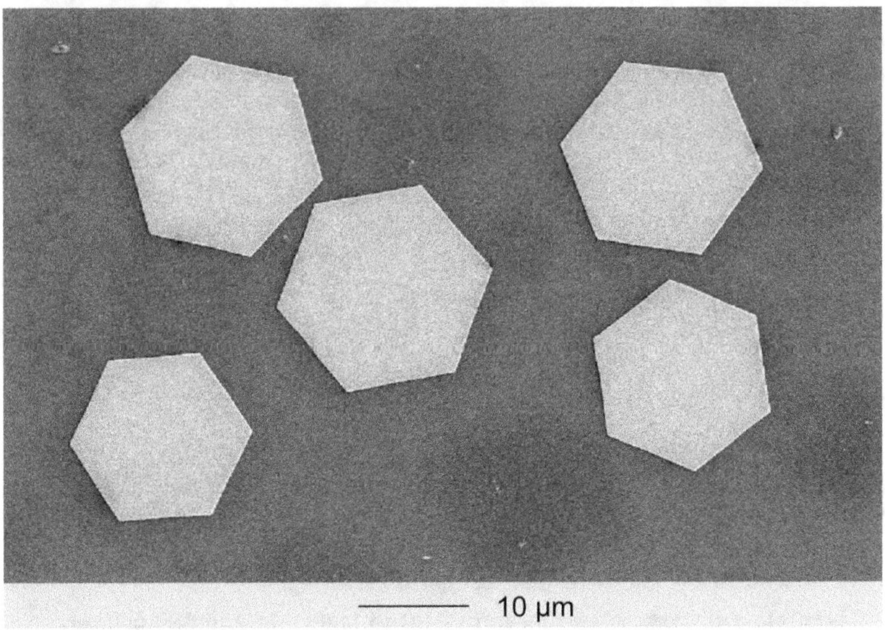

Figure 2.14: Scanning electron micrograph of basic lead carbonate

hydrogenarsenate $PbHAsO_4$, and lead thiosulphate PbS_2O_3, as well as substances such as mercury chloride, manganese and zinc ammoniumphosphate, barium thiosulphate, and barium hypophosphite.

Lacking better alternatives, lead hydrogenphosphate and lead hydrogencarbonate, both of which are chemically inert, played a major role in the production of artificial horn. Occasionally lead thiosulphate was used in special applications as an alternative to basic lead carbonate.

The fully non-toxic pigments based on mica coated with titanium dioxide developed in the last decades meet and surpass all of the advantageous properties of these lead compounds and have consequently replaced them on the market.

2.4 Bismuth oxychloride

In parallel with the monocrystalline pigments based on lead compounds described in Chapter 2.3, bismuth salts such as the basic nitrate $Bi(OH)_2NO_3$ and, in particular, bismuth oxychloride BiOCl have awoken interests for decades [18, 21]. Bismuth oxychloride was the first synthetic non-toxic pearl luster pigment that existed. It is precipitated by hydrolysis of acidic bismuth salt solutions in the presence of chloride ions:

$$Bi^{3+} + Cl^- + 3\ H_2O \rightarrow BiOCl + 2\ H_3O^+$$

Depending upon the precipitation conditions (concentration, temperature, pH value, pressure, addition of surfactants), irregular tinsel or extremely flat octahedra develop (Figure 2.15). The growth of lamellar monocrystals is very difficult to achieve. It depends mainly upon experience and application of a precise experimental procedure. Even the slightest deviation in pH value, temperature or concentration can affect the product quality. Surface-active additives can inhibit the "natural growth" of the crystals leading to extremely flat octahedral, or rectangular, octagonal or even round, lens-shaped BiOCl crystals.

Despite having a very high refractive index of 2.15, bismuth oxychloride does not achieve the brilliance of basic lead carbonate due to its inferior crystal forms. However, its soft luster, good hiding power, and non-toxicity have made BiOCl irreplaceable for decorative cosmetics.

Pure bismuth oxychloride is mainly used in the cosmetic industry and is available in three forms, with varying pearl luster effect depending upon the aspect ratio and crystal size:

- low or medium gloss powder (aspect ratio 1:10 to 1:15), mainly implemented as a white, gloss filler with an excellent skin feeling,
- high gloss dispersion (aspect ratio 1:20 to 1:40), rectangular or hexagonal platelets used in nitro-cellulose lacquers (nail varnish) or castor oil (lipsticks),

- very high gloss dispersion (aspect ratio >1:50), consisting of lens-shaped platelets in nitro-cellulose varnish, castor oil, butyl acetate or other organic systems, depending upon the application area.

The pigments are also available as BiOCl coated mica or talc or a mixture with organic and/or inorganic colorants. These preparations contain 35 to 50 % bismuth oxychloride and are lighter than pure BiOCl (density 7.73 g/cm^3). Their gloss lies between the dispersions and the dry BiOCl powders.

The limiting factors for further applications of these pigments are

- poor light stability (even sunlight can turn the pigments from silver-white to metallic gray),
- relatively high costs,
- quick sedimentation, and
- poor mechanical stability.

It is remarkable for the use of BiOCl pigments in paints, that the single particles are nearly invisible to the human eye in the application medium. As a result, homogeneous surfaces without a sparkle effect can be varnished or printed. It is therefore possible to achieve an appearance very close to the so-called "chromium-mirror-

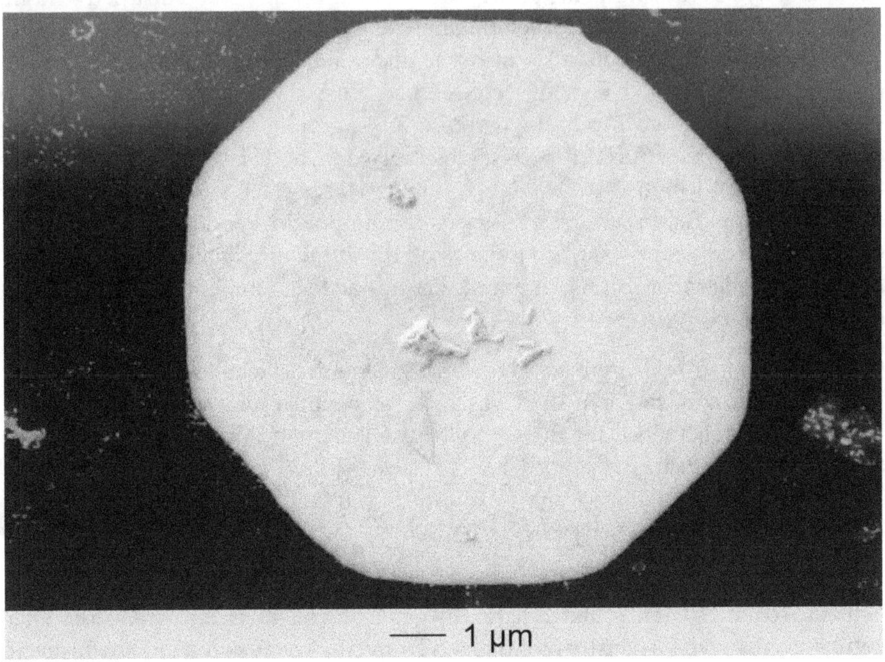

Figure 2.15: Scanning electron micrograph of a bismuth oxychloride crystal

effect" without having problems during the application [22]. The poor light stability is however a disadvantage for outdoor applications. Bismuth oxychloride under exposure to UV light tends not only towards graying, but loses also a part of its luster. Due to the comparatively weak mechanical stability, the pigment particles should not be exposed to stronger shear stresses.

The reactions leading to the graying process by light exposure are not yet fully understood. However, BiOCl types do exist meanwhile showing an improved light stability. High-lustrous bismuth oxychloride pigments are up to now almost only sold in the form of preparations. In the paint and printing industries, special pastes are used (pigment content from 50 to 80 %, adjusted to the paint and printing formulations). These pastes are already partly equipped with suitable UV stabilizers. The recent development of more stable types has led to the situation that bismuth oxychloride pigments are now more and more used also in non-cosmetic systems, especially in industrial coatings and plastics. Such pigments attract attention by creating noticeable luster and iridescence effects in the application system. Strong and saturated color effects, good hiding power and whiteness with a bluish tint can be adjusted in a systematic manner. The specific brilliance is generated by the reflection of the incident light on the optimally formed BiOCl crystals [22].

Bismuth oxychloride pigments possess due to their bright color a distinctive high styling potential. They show excellent application properties in coatings (dispersions are available for both solvent-borne and water-borne paint systems). Pigment particles suitable for such effects exhibit a narrow distribution of diameter especially in the range 7 to 14 µm. Printing inks, graphic-art coatings, industrial paints as well as coatings for metals, plastics, wood, and leather are applications for which BiOCl pigments are highly suited. The pigments are predominantly applied by dipping, spraying, and roll-coating. Important applications in the field of lacquer coating are as surface finishes on furniture, electronic components, or household equipment. BiOCl pigments are utilized also in high-quality printing inks for packaging.

Other customers for bismuth oxychloride pigments are the button and the jewelry industries. The application of BiOCl as an X-ray contrast substance should also be noted. The current market for BiOCl pigments in all applications is estimated to be at approximately 500 t per year [18].

2.5 Micaceous iron oxide

Micaceous iron oxide (iron mica, micaceous iron ore) consists of pure or doped hematite (α-Fe_2O_3). It is found as naturally occurring platelets, e.g. in Austria. This natural product with a density of 4.6 to 4.8 g/cm^3, dark gray color and low luster is nearly exclusively used in corrosion protecting coatings.

Micaceous iron oxide can be synthetically obtained as platelets by hydrothermal reaction in alkaline media. The single-crystalline platelets formed by this reaction show also a color as dark as the natural iron mica. If substantial amounts of doping materials are incorporated, the aspect ratio can be increased up to 100, resulting in much better luster [23]. By this means the luster of the particles is increased significantly. The color can also be shifted from a dull dark brown to a more attractive reddish brown allowing the use of the platelets for decorative applications.

The most important doping constituents are Al_2O_3 and Mn_2O_3. The use of these oxides supports the formation of spinel structures. Incorporation of Al_2O_3 leads to thinner and larger particles whereas the thickness of the platelets is reduced by Mn_2O_3 doping [23]. The use of ZrO_2, B_2O_3, P_2O_5, SiO_2, or additional Al_2O_3 reduces the agglomeration in the application system. Chromates, molybdates, tungstates, or titanates can be applied for the adjustment of various color shades.

Iron(II) sulfate is used as starting material for the synthesis of micaceous iron oxide. Iron(II) is oxidized with air to iron(III) in an agitated vessel. The iron(III) so formed is precipitated as FeOOH by the alkali hydroxide present in the vessel [24, 25].

$$2\ FeSO_4 + 4\ NaOH + O_2 \rightarrow 2\ FeOOH + 2\ Na_2SO_4 + 2\ H_2O$$

In the presence of aluminum oxide and potassium permanganate, the FeOOH suspension is converted to platelet-shaped Al and Mn containing iron(III) oxide. The reaction takes place in an autoclave equipped with stirrer and heating [24, 25]. The size of the platelets is dictated by control of the parameters temperature, pressure, reaction time, and concentration of alkali, FeOOH and doping agents.

$$2\ FeOOH + (Al_2O_3,\ NaOH,\ KMnO_4) \rightarrow Al_xMn_yFe_{2-x-y}O_3 + H_2O$$

The FeOOH suspension is heated up together with the doping agents to temperatures above 170 °C, typically to 250 to 300 °C. Platelets of doped iron oxides are formed mostly using reaction times of several hours and pressures of $3 \cdot 10^7$ Pa. In a second step of the reaction, the pH value is further increased in order to achieve platelets in the required final size and shape [24, 25]. Platelet-shaped iron oxide is preferably used as effect pigment in paints and coatings.

2.6 Titanium dioxide flakes

Titanium dioxide platelets (TiO_2-flakes) can be produced by breaking up continuous films of TiO_2. Such films can effectively be obtained using a web-coating process where $TiOCl_2$ is thermally hydrolyzed on the surface of the web [26–28].

TiO_2-flakes can also be achieved from TiO_2-mica pigments by dissolving the substrate in strong acids or hydroxides [29–33]. The TiO_2-coating of the starting pigment remains as insoluble TiO_2-platelets in the suspension.

The titanium dioxide flakes obtained in this way are not single crystals but polycrystalline and slightly porous. They show limited mechanical stability in most cases and can, therefore, not be used in technical applications where stress is exerted. Applications in cosmetic formulations could be feasible under suitable conditions.

As effect pigments, TiO_2-flakes show typical interference colors, which are dependent on the respective thickness of the platelets [18]. Further variations of color and luster of these effect pigments can be achieved by coating the platelets with different metal oxides in single and multi-layers [27, 28].

2.7 Flaked organic effect pigments

Some organic pigments, which are typically used as colored pigments, can also be forced to crystallize in the form of thin flakes. Typical examples are 1,4-diketo-3,6-diaryl-pyrrolo(3,4-c)-pyrrole (DPP) [34, 35], 2,9-dichlorochinacridone [36], and metal phthalocyanines [37]. However, the difference between their refractive index and that of the typical application media is insufficient to generate strong interference colors and luster effects [18]. In most cases, the aspect ratio of these crystals is also much smaller than that of the inorganic effect pigments described in previous chapters. The technical importance of platelet-shaped organic effect pigments is comparatively small due to their limited optical properties.

2.8 Metal oxide mica pigments

Possessing a very high refractive index (2.5 to 2.7) and chemical and thermal stability, titanium dioxide is the most important white pigment on the market today. Additional advantages can be found in its light stability, low price, and non-toxicity. A mono-crystalline, lamellar titanium dioxide undoubtedly would be a very interesting effect pigment. Unfortunately, all attempts to grow titanium dioxide in a lamellar form have failed. The main reason for this is the tendency of the oxide to crystallize as tetragonal bipyramids. As described in Chapter 2.6, the properties of polycrystalline TiO_2-platelets are insufficient to fulfill the high demands of the application fields for effect pigments.

Coating concepts

An alternative can be found in coating titanium dioxide on the surface of a suitable carrier, i.e. a lamellar substrate. Such a substrate needs to be transparent, chemically inert, thermally and mechanically stable, and last but not least, cheap and easy to obtain. Commonly occurring mineral mica meets these requirements almost completely. It can be ground easily into thin platelets of the desired diameter and thickness [2, 4, 18, 21, 38].

The most important class of special effect pigments today consists of natural mica platelets coated with thin films of transparent metal oxides. Titanium dioxide plays a particularly important role in this respect, but other oxides such as iron(III) oxide, tin dioxide, and chromium(III) oxide must also be mentioned. Mica is an alumosilicate with a layer structure (Figure 2.16).

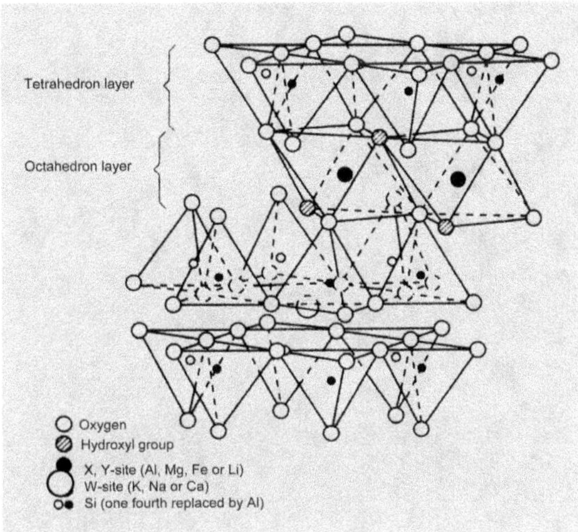

Figure 2.16: Three-dimensional structure of muscovite mica

The mica structure consists of inter-connected tetrahedral and octahedral layers. The tetrahedral layers consist of SiO_4 and AlO_4 tetrahedrons in a ratio of 3 : 1 which are linked with layers consisting of octahedral AlO_6-units (with Al in the octahedrons partly replaced by Mg, Fe, or Li). Further tetrahedral layers follow the octahedral layers in a defined sequence. There is a comparatively large separation in the mica structure between the tetrahedral and octahedral layers. Within the structure Na, K, or Ca ions are located in interstitial positions. In addition, the interstices are the weakest areas of mechanical strength in the crystal structure. They act therefore as cleavage planes when the mica is put under mechanical strain, e.g. during grinding. Ultimately the interstices are the reason why it is possible to cleave large mica blocks into small and thin platelets. These platelets are suitable for coating with thin metal oxide layers to form effect pigments.

Transparent muscovite is commonly used for effect pigments, but in some cases synthetically manufactured phlogopite is chosen as the carrier. Muscovite is found throughout the whole world, but only a few deposits are suitable as starting material for metal oxide mica pigments. Muscovite has been mainly used as cheap filler [39, 40].

Pigment particle dimensions

Selection and preparation of the mica are the most important steps that determine the quality of the final effect pigments. The ratio of diameter to thickness (aspect ratio) of the pigment particles is directly proportional to the particle size distribution of the starting mica platelets. The mica particles are usually between 300 and 600 nm thick. The optimum diameter varies greatly depending upon the selected application.

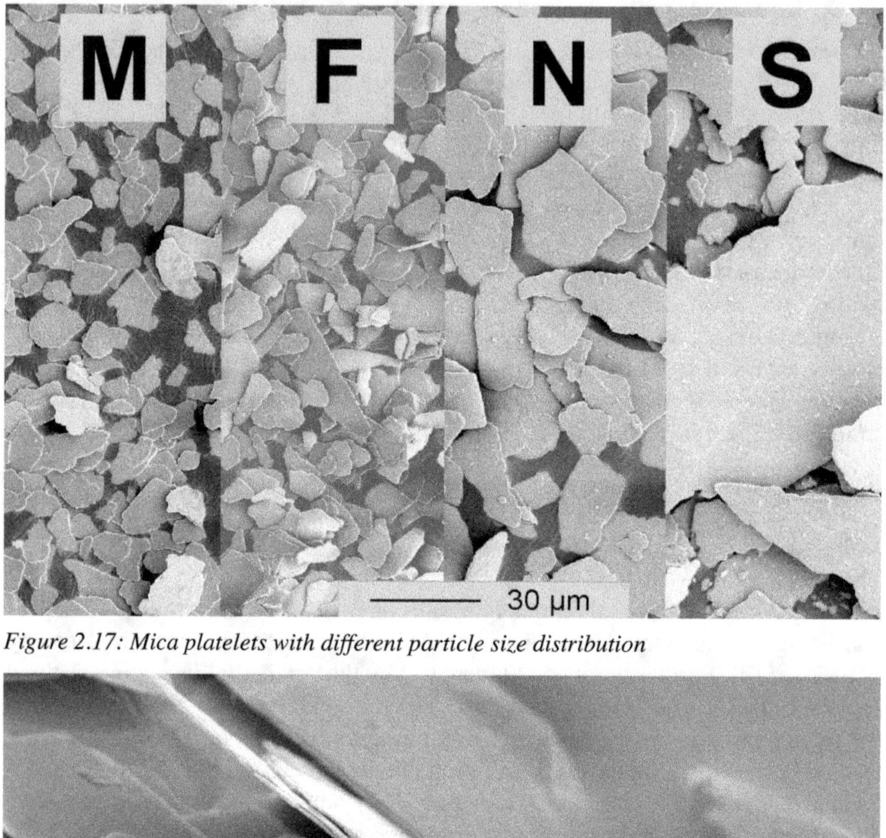

Figure 2.17: Mica platelets with different particle size distribution

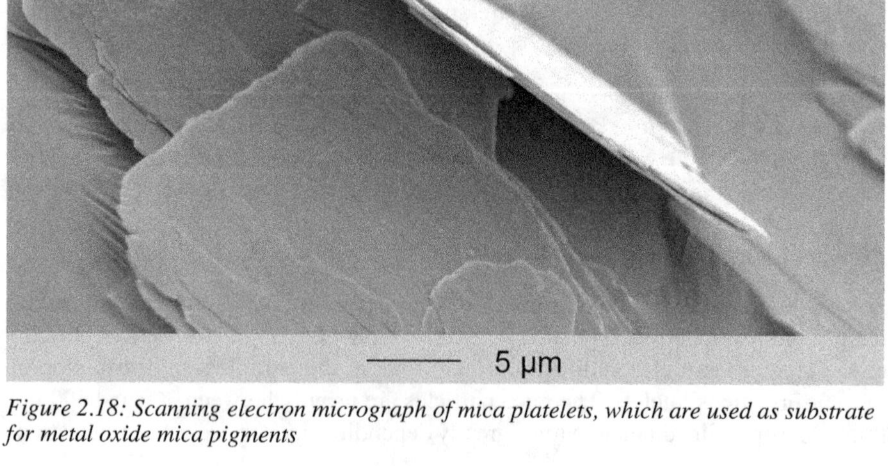

Figure 2.18: Scanning electron micrograph of mica platelets, which are used as substrate for metal oxide mica pigments

A frequently used classification distinguishes five classes of the particle size distribution (Figure 2.17):

- M: 1 to 15 µm
- F: 5 to 20 µm
- N: 10 to 50 µm
- S: 10 to 130 µm
- L: 40 to 200 µm

Figure 2.19: Atomic force micrograph of the surface of a mica platelet (muscovite) after the grinding procedure

Luster and hiding power of the effect pigments in the application system are the two most important properties influenced by the particle size distribution of the chosen mica. There is a general rule as follows, larger mica particle diameters lead to increased luster and smaller mica particle diameters generate increased hiding power.

In summary:

- **large particles** = large smooth surfaces, uniform reflection of incident light leading to excellent luster and brilliance,
- **small particles** = smaller flat surfaces, more edges and corners, increased scattering and refraction, therefore better hiding power.

Figure 2.18 shows a Scanning Electron Micrograph (SEM picture) of mica platelets as they are produced by cleaving muscovite mica. It can be seen that the resulting platelets are not uniform in thickness. They contain steps, which result from the grinding process, but overall they possess a very smooth surface. This is also proven by Atomic Force Microscopy (AFM) of thin mica platelets, which are produced by the cleavage of rough mica blocks (Figure 2.19).

Structural principles of metal oxide mica pigments

In the most simple case, a metal oxide mica pigment consists of three layers with different refractive indices n_1, n_2, n_3; where $n_1 = n_3$ (Figure 2.20, page 44).

There are four interfaces in this case (P_1 to P_4):

- P_1 and P_4: surrounding medium/metal oxide,
- P_2 and P_3: metal oxide/mica.

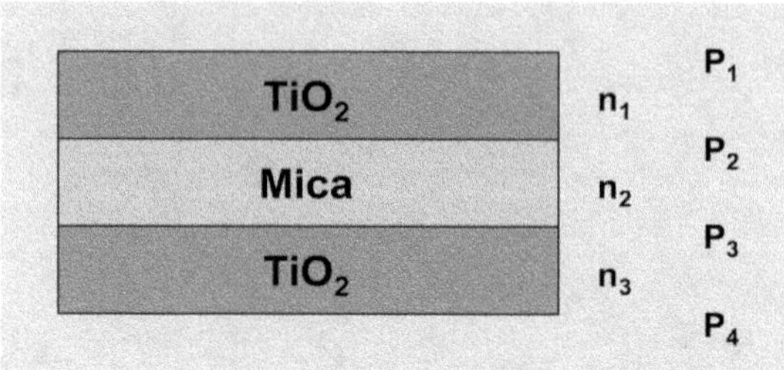

Figure 2.20: Schematic structure of a titanium dioxide mica pigment with the four existing interfaces (P_1–P_4)

When examining a titanium dioxide mica pigment, it is easy to recognize that the interference phenomena take place by reflection on all six combinations of interfaces. Some combinations such as P_1P_2 and P_3P_4 or P_1P_3 and P_2P_4 create identical interference phenomena. The average thickness of the mica platelets is close to a mean statistical value. The interference effects resulting from the interface between metal oxide coating and mica substrate (P_2, P_3) combine together to create a background white reflection. The interference color resulting from a large number of pigment particles only depends upon the thickness of the upper and lower metal oxide layer [41–43].

The development of mica pigments began with TiO_2-types. The earliest types of pearl luster pigments were silverwhite. As the TiO_2-layer thickness is so thin in this case, there is no interference effect present. Figure 2.21 gives an overview of the important metal oxide mica pigments subsequently developed. With increasing TiO_2 layer thickness on the mica, an interference color develops, varying from gold to green.

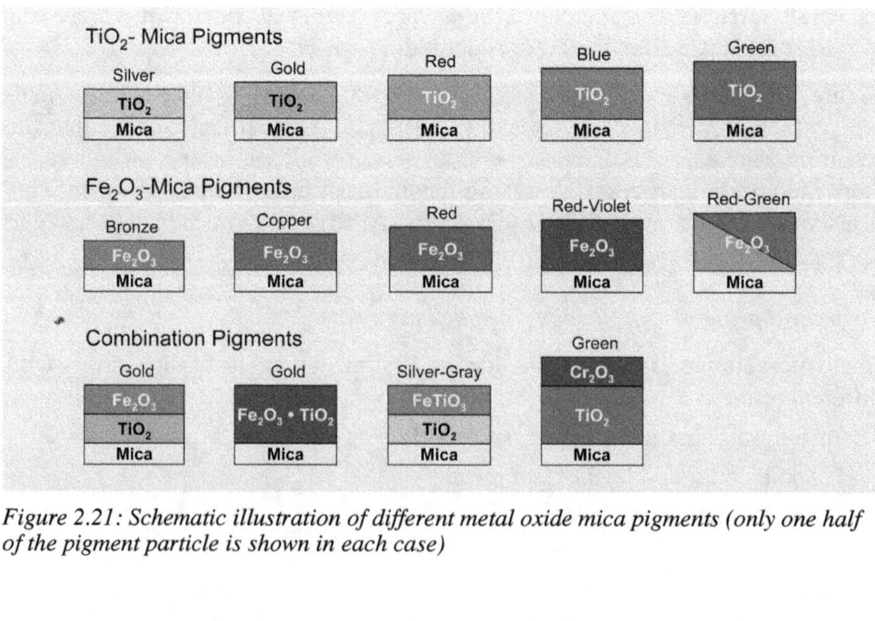

Figure 2.21: Schematic illustration of different metal oxide mica pigments (only one half of the pigment particle is shown in each case)

Brilliant combination pigments were introduced later to the market. They included products containing TiO_2, Fe_2O_3, Cr_2O_3 and other pigments and also dyes in the layer on the substrate. The colors arising develop due to a combination of base colors and interference colors. Depending upon the composition and viewing angle, the pigment may have one color (interference color similar to the base color) or two colors (interference color and base color are different). The synthesis of mica pigments with a transparent iron(III) oxide layer started in the late 1970's.

Manufacturing process

Before manufacturing metal oxide mica pigments, the natural mica must be prepared. The rough mica blocks are ground and then classified into a range of particle size distributions. The subsequent coating step uses a wet chemical process to precipitate metal hydroxides or oxide hydrates on the surface of the fine mica particles. After the coating, filtration, drying and precise calcination processes follow. All these steps are of utmost importance for the quality of the pigments. The complete manufacturing process is shown schematically in Figure 2.22.

Numerous pigments need an additional surface treatment with transparent and colorless oxide films. These layers provide TiO_2-mica pigments with an improved light stability, for example, by reducing the photo activity of the TiO_2-surface [2,18,43]. In addition, the interaction between pigment and application medium can be optimized. The latter point is equally valid for pigments with iron oxide or chromium oxide surfaces.

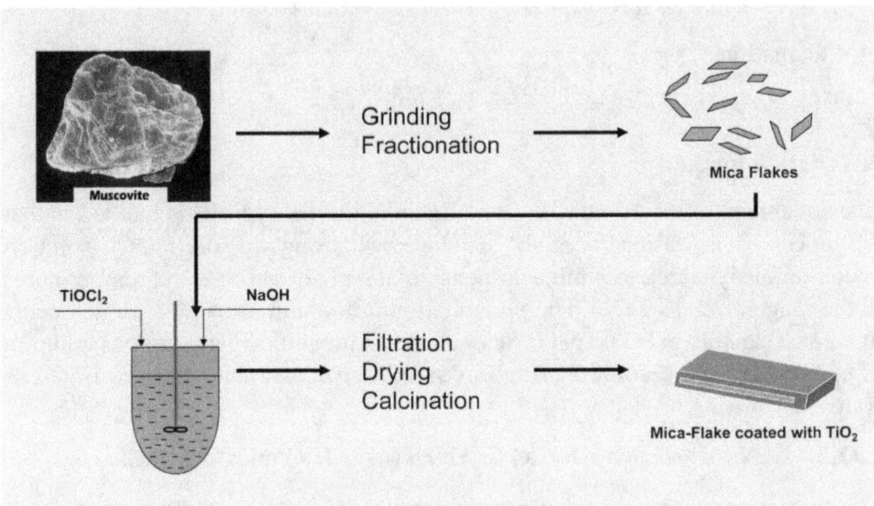

Figure 2.22: Process scheme for metal oxide mica pigments

2.8.1 Titanium dioxide mica pigments

The era of metal oxide mica pigments began with products based on TiO_2-coatings on a muscovite substrate [2,4,18,21,43]. The coating of TiO_2 on mica is usually carried out by the hydrolysis of titanium salt solutions in an aqueous mica suspension. The first synthesis of TiO_2-mica pigments was published as early as 1942 [44]. Commercial use was started first by DuPont in the early 1960's [45,46]. Further developments were mainly driven by Merck [47] and Mearl [48]. In these products, mica acts as a template for the deposited metal oxide layers during the synthesis and as a mechanical support for the thin optical layers formed in the final pigment.

To date, two manufacturing methods have been established:

1. Homogeneous hydrolysis

A measured amount of titanium sulphate is added to an aqueous mica suspension. By slowly heating the suspension the titanium salt is hydrolyzed to insoluble titanium dioxide hydrate, which steadily deposits as a homogeneous coating on the mica particles. After all of the titanium salt has been hydrolyzed, the coated particles are filtered off, dried and finally calcined at 700 to 900 °C in a suitable kiln. At these high temperatures, water is eliminated and a thin compact coating of titanium dioxide is formed around the mica. Equations for the main reactions are as follows:

a) Coating

$$TiOSO_4 + \text{mica} + n\,H_2O \rightarrow TiO_2 \cdot (n\text{-}1)\,H_2O/\text{mica} + H_2SO_4$$

b) Calcination

$$TiO_2 \cdot (n\text{-}1)\,H_2O/\text{mica} \xrightarrow{-(n-1)H_2O} TiO_2/\text{mica}$$

2. Titration method

The advantage of this method in comparison with the hydrolysis method is that the processes are easier to control. An aqueous, strongly acidic $TiOCl_2$-solution is continuously added to a mica suspension at a pH of about 2 and temperatures in the range of 60 to 90 °C. The precipitation of titanium dioxide hydrate occurs at a precise acidic pH. The pH is held at a constant value by continuous addition of sodium hydroxide solution. A highly simplified equation for this reaction is as follows:

$$TiOCl_2 + 2\,NaOH + \text{mica} + n\,H_2O \rightarrow TiO_2 \cdot (n+1)\,H_2O/\text{mica} + 2\,NaCl$$

The further processing of the pigment particles is identical to the homogeneous hydrolysis method.

Structure and distinguishing details

Titanium dioxide deposited on mica platelets by precipitation at pH values above 1.5 is formed in the anatase modification alone. Even with annealing at temperatures around 1000 °C, no rutile phase can be detected in the TiO_2-layers. On the other hand, titanium dioxide precipitated in an aqueous solution without carrier particles, under suitable conditions can completely crystallize in the rutile modification at temperatures as low as 700 °C [49]. The use of phlogopite mica instead of muscovite mica can force rutile formation, which can be observed in this case at temperatures above 850 °C [50]. The reason for this difference in behavior is due to the migration of Al ions out of the muscovite mica into the anatase layer leading to inhibition of the phase transformation [50–53]. On the other hand, the anatase modification is formed below 1000 °C only when Al-free silicon dioxide or aluminum oxide in the form of thin platelets are used for the deposition of TiO_2. The most likely explanation of this is due to the considerable mechanical strain in the TiO_2-layer, which counteracts the volume reduction arising in rutile formation and therefore inhibits the phase transformation [49, 54].

The anatase layers on the mica platelets possess a preferred crystal orientation when they are very thin, e.g. below 40 nm [50, 54]. This preferred orientation is lost during further growth of the layer thickness. Moreover, the uncalcined TiO_2-layers show a deviation of the lattice parameters in comparison to anatase in the compact and dense solid form (single crystal/pressed pellet). The differences are reduced after the annealing step [54]. These phenomena are evidences of the strong template effect of muscovite mica when used as a substrate during the coating and calcination processes for the pigment manufacture.

The rutile modification, in comparison, has advantages such as a higher refractive index and better light stability. Consequently, the possibility to produce material in the rutile forms was examined. It was discovered that pre-coating the mica with small amounts of tin dioxide drives the reaction in the desired direction:

$$SnCl_4 + 4\,NaOH + mica \rightarrow SnO_2/mica + 2\,H_2O + 4\,NaCl$$

$$SnO_2/mica + TiOCl_2 + 2\,NaOH \rightarrow TiO_2/SnO_2/mica + 2\,NaCl + H_2O$$

$$TiO_2/SnO_2/mica \xrightarrow{>700\,°C} TiO_2\text{ (rutile)}/SnO_2/mica$$

Tin dioxide crystallizes with a rutile structure and forces the precipitated titanium dioxide to form this phase as well.

Figure 2.23 (left) shows the SEM picture of a TiO_2-mica pigment. Both the shape and size of the individual pigment particles can clearly be seen. Figure 2.23 (right) and Figure 2.24 (left) show cross-sections of individual pigment particles. The crystalline titanium dioxide layer formed during coating and calcination can be clearly seen on both sides of the mica platelet. This TiO_2-layer covers the substrate

Figure 2.23: Scanning electron micrographs of a titanium dioxide mica pigment. Left: Overview micrograph. Right: Cross-section through one particle. Layer structure: TiO_2-mica-TiO_2.

platelets completely. Its thickness can be adjusted with nanometer precision during the manufacturing process. The exact control of this layer thickness is essential for the manufacture of special effect pigments with reproducible color properties. This is especially true for instance, for the silver white and interference pigments described in the following chapters.

Titanium dioxide mica pigments are often divided into silver white pearl luster pigments and interference pigments. An aqueous mica suspension does not have great appeal before the addition of the titanium salt solution and its hydrolysis. There is no luster whatsoever, which is not surprising because the refractive indices of mica and water are not quite similar. When, however, a titanium salt is hydrolyzed according to one of the methods above, the suspension appears at first silky-matt and as the reaction continues a more intensive luster develops. If the reaction is stopped at this point and the particles are drawn off, silver white luster pigments are obtained. The pigments have a titanium dioxide coating that is about 40 to 60 nm thick.

The pigments possess a neutral silver white luster and form the most important group of pearl luster pigments. Depending upon the particle size distribution, they exhibit silky-matt, brilliant or glittering luster. Combined with other transparent pigments, they can impart a color shade with a silvery, metallic luster. They are often used, however, without additional colorants. If a little carbon black is added, a silver-gray effect can be obtained – similar to that of aluminum effect pigments.

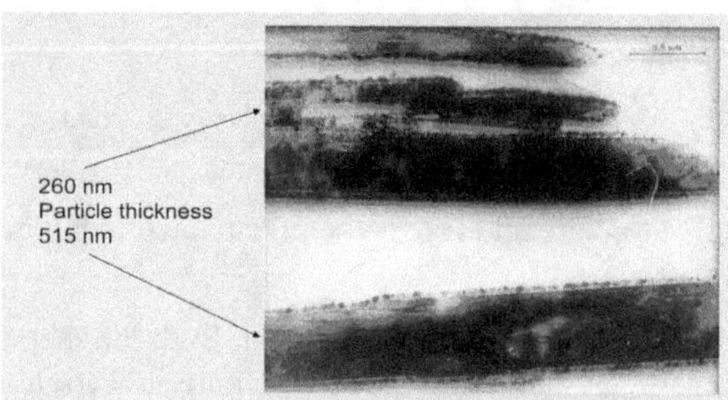

Figure 2.24: Transmission electron micrograph of a titanium dioxide mica pigment (cross-section with indication of the particle thickness)

Interference colors

An interesting optical phenomenon occurs during the titration process when the coating of mica with titanium dioxide is not stopped at the silver-white stage. If further titanium dioxide is added to the reaction mixture, the pigment particles start to shine gold-yellow without losing their luster. After the addition of further titanium dioxide, this color disappears to be replaced by copper-red, and then lilac. This is followed by a lustrous blue, which then becomes turquoise and finally green. If still more titanium dioxide is added, this fascinating color display starts from the beginning again.

What is the explanation for this phenomenon? The titanium dioxide layers have reached a thickness where they behave according to the optical principles of thin platelets light interference occurs. Figure 2.25 demonstrates how the layer thickness of TiO_2 determines the interference color. Figure 2.26 shows from another angle of view the dependence of the interference color on the TiO_2 layer thickness.

The color values in the illustration were calculated using the equations for the optics of thin films and are expressed here by the L*a*b*-values according to Hunter [2, 43]. In this system the +a-value represents red, the –a-value corresponds to its complementary color green. Yellow colors are displayed as +b-values and blue colors are displayed as –b-values (Figure 2.27). By superposition of the a- and b-curves, one can calculate the TiO_2-layer thickness needed to obtain a specific color.

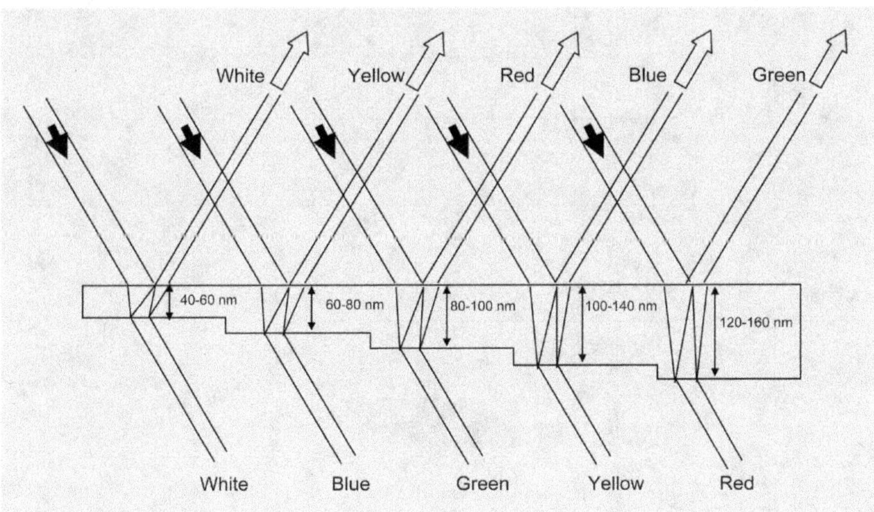

Figure 2.25: The thickness of the titanium dioxide layer determines the interference color

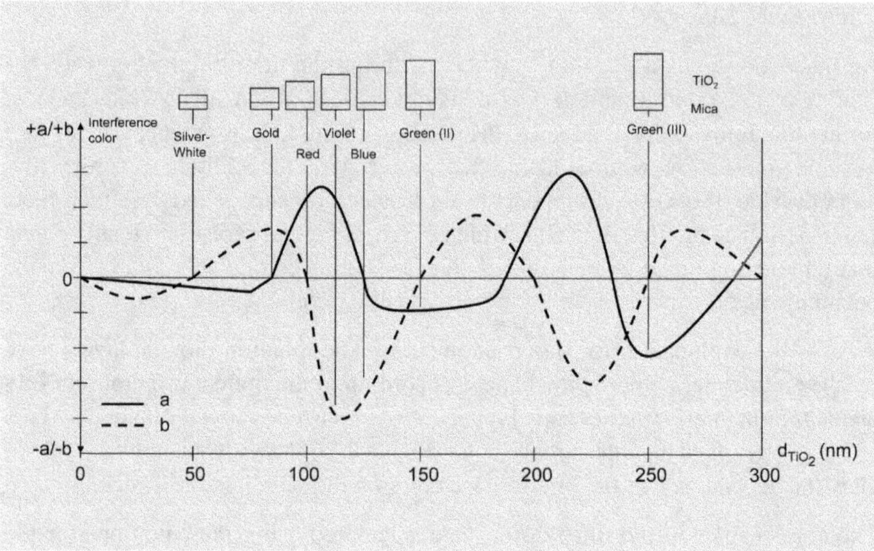

Figure 2.26: TiO_2-mica pigments: The dependence of interference color according to titanium dioxide layer thickness.

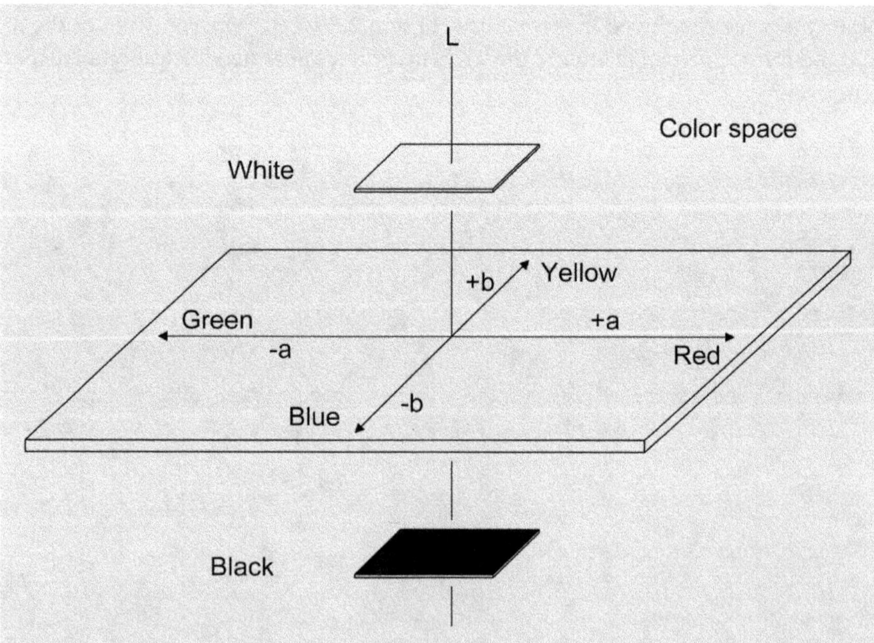

Figure 2.27 L,a,b-color system according to Hunter

If a red interference color is required, for example, one orients at first toward the maximum for the +a-value, which lies at a TiO_2-layer thickness of 110 nm (Figure 2.26). To avoid false conclusions, the corresponding –b-value (blue value), which is also very high at 110 nm, should simultaneously be observed.

To obtain a pure interference red, a layer thickness should be selected which lies beneath the maximum value for +a, and where the –b-value is at a minimum. The purest red, for example, can be obtained with a TiO_2-layer thickness of 100 nm. At 110 nm, the combination of the high +a-value and the high –b-value produce a violet shade.

In Figure 2.26, it is easy to see that one specific interference color can be achieved with different TiO_2-layer thicknesses. This can be illustrated with interference green when observing the curve from left to right, that is with increasing layer thickness. First order green is so weak and overpowered by the intensity of blue and yellow that it is not worth manufacturing and selling. The first pure green, second order green, develops at a TiO_2-layer thickness of 150 nm. It is frequently used today as a green interference pigment.

If the a-curve is followed even further, another pure green is seen at a layer thickness of 250 nm, the third order green, with an intensity even greater than second order green. However, Figure 2.26 shows that the other interference colors of higher order have no obvious advantage versus the corresponding colors of lower order.

The exact adjustment of the TiO_2-layer thickness in the nanometer range mentioned above, is basically the decisive factor for the achievement of a defined interference color. Even thickness deviations of only a few nanometers lead to other colors and effects visible to the human eye. Thus, one can truly describe the manufacture of metal oxide mica pigments as nano-technology, whereby not nano-particles but nano-layers are the result of the activities [55].

Iridescent effects

Interference pigments show a significant color flop (color changing effect, flip-flop effect). The color varies with the angle from which the pigment is observed. The interference color can only be seen from a certain angle, the glancing angle. The color that can be seen from the other positions depends on several factors.

Let us presume that an interference blue pigment is printed onto a white, a black, and a colored, e.g., a green background (Figure 2.28). Interference, as mentioned in the description of light, can be considered as selective reflection of light rays. For the interaction of light with the blue pigment, the following statements can be made:

a) White background
Incident white light is reflected at the glancing angle as blue and the remaining light passes through the transparent pigment. This part of the light is complementary to blue therefore it is yellow. The transmitted yellow light is diffusely scattered in all directions from the white background. At the glancing angle, the intense interference color covers the yellow light, which is diffusely scattered by the background. From all other directions the object appears pale yellow.

b) Black background
Black absorbs nearly all the light, which is in contact with it, including the transmitted yellow light that makes its way through the interference blue pigment. At the glancing angle, the eye again sees a lustrous blue object. Since light is scattered at the corners and edges of the pigment particles, the surface of the object when seen from other angles does not appear totally black, but more or less gray-blue.

c) Colored background
Due to the transparency of the pearl luster pigments the background color shines through. At the glancing angle though, the background color is covered by the interference color. Interference color and background color can be identical or different, which makes all sorts of flip-flop effects possible. Particularly interesting effects can be achieved on curved surfaces, where both colors can be seen at the same time. When complementary-colored interference pigments are combined, due to the additive color mixture, white is obtained. Mixtures of pigments from neighboring colors lead to a lighter shade with a loss of color brilliance.

Economic aspects

Mica cannot be ground to extremely thin platelets. Since the mica flakes are coated on both sides with a TiO_2-layer, the particles of the titanium dioxide mica pigments

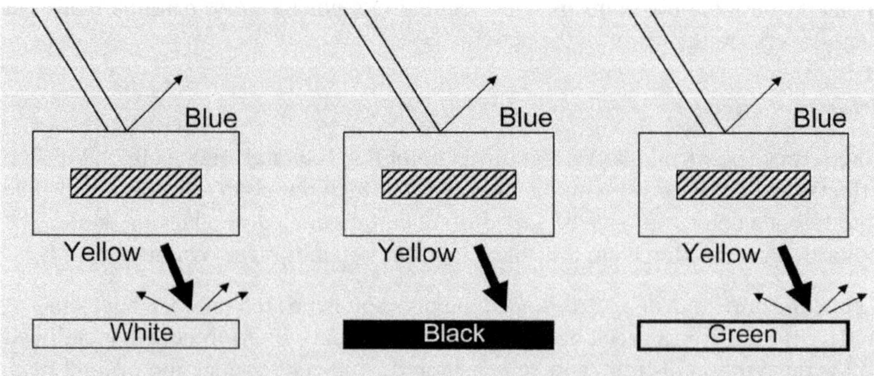

Figure 2.28: Interference pigments upon white (left), black (middle) and green (right) background

are approximately 5 to 10 times as thick as particles of basic lead carbonate. If we consider that the TiO_2-coating is optically separated by the mica layer and operates as two parallel reflecting layers, that the high refractive index of TiO_2 leads to a roughly 60 % higher degree of reflection, and finally, that the density of mica is only half that of basic lead carbonate, then an almost equivalent usage factor results for both pigments.

A comparative overview of titanium dioxide mica pigments, basic lead carbonate, natural fish silver and bismuth oxychloride is made in Table 2.2 and the corresponding characteristics in Table 2.3. The obvious advantages of titanium dioxide mica pigments predict a growing demand for this pigment class in the future. The search for manufacturing routes as alternatives to the wet chemical reactions is still being pursued. The gas phase precipitation by chemical vapor deposition (CVD) in a fluidized bed reactor is amongst these alternatives [23, 56, 57].

The manufacture of titanium dioxide mica pigments using the CVD-process is, however, so difficult, that to date it has not achieved technical importance. The process is performed basically in such a manner, that $TiCl_4$ is converted in the gaseous phase with water vapor. Steadily swirled mica platelets are mixed together

Table 2.2: Properties and application of pearl luster pigments

Pearl luster pigment	Advantages	Disadvantages	Major application fields
natural fish silver	very low density high luster non-toxic light stable	high price low hiding power limited availability	nail lacquers
basic lead carbonate	very high luster good hiding power low price light stable	high density limited chemical and thermal stability toxic	buttons bijouterie
bismuth oxychloride	very high luster good hiding power non-toxic	limited light stability high density	decorative cosmetics buttons bijouterie
titanium dioxide-mica	high luster good hiding power (dependent on particle size) highest thermal, chemical and mechanical stability non-toxic low price low density	inferior luster in comparison with top qualities of basic lead carbonate and bismuth oxychloride	decorative and skincare cosmetics automotive and industrial coatings plastics printing inks ceramic products

Table 2.3: Technical data of pearl luster pigments

Pearl luster pigment	Shape	Particle size	Thickness	Density (g/cm³)
natural fish silver	needles, longish platelets	10 to 40 µm	40 to 50 nm	1.6
basic lead carbonate	hexagonal crystals	4 to 20 µm	40 to 70 nm	6.4
bismuth oxychloride	flat tetragonal bipyramidal crystals	5 to 30 µm	100 to 700 nm	7.7
titanium dioxide mica	platelets	1 to 200 µm	200 to 500 nm	3

with the reactants in a fluidized bed apparatus at temperatures above 100 °C leading to the direct formation of TiO_2 layers on the surface of the mica particles:

$$TiCl_4 + 2\,H_2O + \text{mica} \xrightarrow{>100\,°C} TiO_2/\text{mica} + 4\,HCl$$

2.8.2 Iron(III) oxide mica pigments

Following the production of titanium dioxide mica pigments, the coating of mica platelets with Fe_2O_3 soon awoke interest in pigment designers. Iron oxide in the hematite modification (α-Fe_2O_3), with a refractive index of 2.9, good hiding power, and adequate chemical and thermal stability, is particularly suited as an effect pigment.

Manufacturing and properties

Coating of mica with transparent hematite layers follows a wet chemical process similar to that employed for titanium dioxide mica pigments. An iron(II) or iron(III) salt solution is added drop-wise to a mica suspension at a constant pH (analogous to a volumetric titration). The hydrated iron oxide so formed is then dehydrated at temperatures between 700 °C and 900 °C. The following reactions are therefore used:

$$Fe^{3+} + \text{mica} + 5\,H_2O \rightarrow FeO(OH)/\text{mica} + 3\,H_3O^+$$

$$2\,FeO(OH)/\text{mica} \xrightarrow{-H_2O} Fe_2O_3/\text{mica}$$

The iron oxide layers prepared consist mainly of hematite at this stage. In a few cases only, small amounts of goethite (α-FeO(OH)) can be detected. The nanocrystallites formed possess diameters of 20 to 40 nm and show a strongly preferred orientation, which has its origin in a template effect arising from the muscovite mica. The granular layer structure is sintered during annealing at 700 to 900 °C to form a less porous, apparently glass-like, but crystalline, coherent phase [54]. Doping with borates and phosphates can hinder or reduce the sintering.

The color and the luster effect of the pigments made using this method depend upon two optical properties: absorption and interference. The layer thickness of iron oxide can be controlled in the manufacturing process so that the red or red-brown absorption color of hematite is superimposed by an interference color. Bronze, copper, and various shades of red are the result. Using relatively thick layers, the manufacture of red-violet or even red-green iron oxide mica pigments is possible. A layer thickness of 75 nm Fe_2O_3, for example, leads to a bronze color and a thickness of 85 nm gives a red shade. In the case of bronze, copper, and red shades, the red-brown absorption color of the pigments is enhanced by practically identical interference colors (yellow, copper, red). This ensures very intense colors at the glancing angle. The red-violet and red-green pigments, due to a thicker iron oxide layer, have a violet or green interference color at the glancing angle in addition to the red-brown absorption color of the hematite.

Figure 2.29 illustrates the combination of light absorption and interference with iron oxide mica pigments based on calculated a- and b-curves. If one follows the course of the curves, the layer thickness of Fe_2O_3 for a particular interference color can be predicted. The thinnest layer, showing a strong golden interference color, creates a bronze-colored pigment due to the combination with the absorption color. The layer thickness with identical +a (red) and +b (yellow) values creates a copper-colored pigment.

The development of the a- and b-curves towards higher values of iron oxide layer thickness shows that a strong red is just as easy to produce as a violet or a red with intense green interference color.

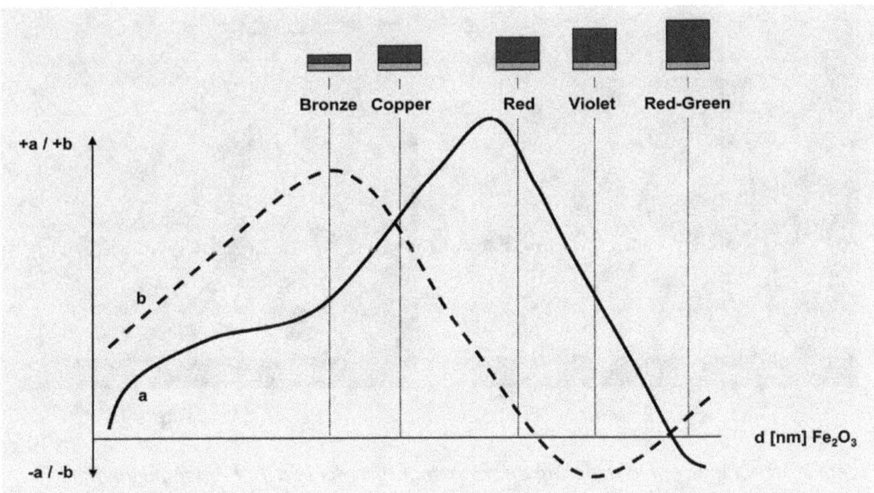

Figure 2.29: Dependence of the a,b-color values from the iron oxide layer thickness in iron oxide mica pigments

The adjustment of defined Fe_2O_3-layer thicknesses in the nanometer range is essential for the pigment production as it was already shown for the TiO_2-mica pigments. Even very small deviations of the iron oxide layer thickness lead to notable differences in color and luster effect.

Special characteristics

At a magnification of 400-times under a light microscope, one can see irregularly formed platelets with a relatively broad particle size distribution. The basic color of the pigment is red. However, on careful observation, one can detect different colors within a single platelet. This is due to the uneven thickness of the mica substrate and leads to slightly different layer thicknesses in the iron oxide coating and consequently to different interference colors.

The electron micrographs in Figure 2.30 show a comparison of the surfaces of a titanium dioxide mica pigment and an iron oxide mica pigment. The titanium dioxide layers show a uniform, fine-grained structure formed during a sintering process at high temperatures. The Fe_2O_3-layers, on the other hand, show a very even, sintered surface, where a defined grain structure no longer exists. However, both surface structures lead in a similar manner to the desired luster, interference and iridescence effects.

Iron oxide mica pigments are not photoactive and therefore possess excellent light stability and high brilliance due to their higher refractive index. Furthermore, the

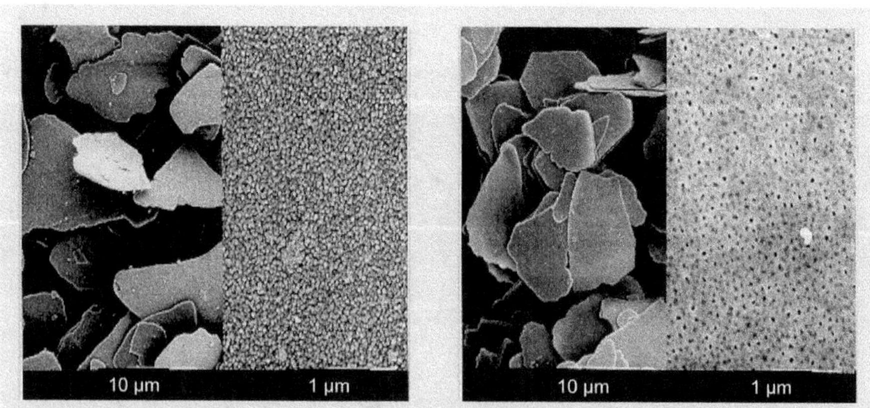

Figure 2.30: Comparative scanning electron micrographs of a TiO_2-mica pigment and a Fe_2O_3-mica pigment. Left in each case overview micrographs, right detail micrographs of the surfaces of single particles. The TiO_2-layer exhibits a fine-grained structure. The Fe_2O_3-layer, on the other hand, shows a sintered surface with larger pores.

red absorption color gives comparatively better hiding power, which can be a decisive advantage in certain applications.

As in the case of titanium dioxide mica pigments, there are now also other manufacturing methods for iron oxide based products, such as the CVD-process. One possible route is to oxidize iron pentacarbonyl in the gas phase and simultaneously deposit the resulting iron oxide onto mica or other platelet-shaped substances [23, 24].

The mica platelets are fluidized in nitrogen gas at temperatures of about 450 °C. Then the reactants $Fe(CO)_5$ and oxygen are injected into the fluidized bed. They must be highly diluted in the inert gas to achieve a suitable coating. The thickness of the iron oxide layer formed on the mica is controlled by the reaction time. Carbon monoxide is formed as a byproduct, which can be catalytically oxidized to carbon dioxide [23]:

$$2\ Fe(CO)_5 + 1.5\ O_2 + \text{mica} \rightarrow Fe_2O_3/\text{mica} + 10\ CO$$

2.8.3 Combination pigments

The combination of transparent absorbing materials with pearlescent pigments provides new color effects not possible by simply mixing these two types of pigment. The effects result from an interaction between absorption and interference when pure titanium dioxide mica pigments are coated with an additional layer of an absorbing colorant (pigment, dye). Of course, one can simply mix interference and absorption pigments to obtain any desired color combination.

Structure and functional principles

In such mixtures, the interference color of the pearl luster pigment dominates at the glancing angle, whereas the color of the absorption pigment can be seen from all other angles. If, however, an extremely thin layer of the absorption pigment in the form of nano-particles is coated as an additional layer onto the interference pigment, the angle-dependent color change becomes more distinct and with greater brilliance. A further advantage of this approach is that it avoids dispersion problems, which can arise when using transparent absorption pigments due to their small particle size and correspondingly large specific surface areas [58].

The structure of a combination pigment is schematically shown in Figure 2.31, where inorganic and organic colorants are coated onto the surface of the titanium dioxide layer. The layer thickness of the titanium dioxide determines the brilliance and the interference effect at glancing angle, while the effect of the transparent colorant dominates from all other angles. If the colorant and the interference pigment have an identical color, a strongly colored effect pigment results with a luster flop that can be seen at all angles. An example of this is the coating of interference blue with Prussian blue.

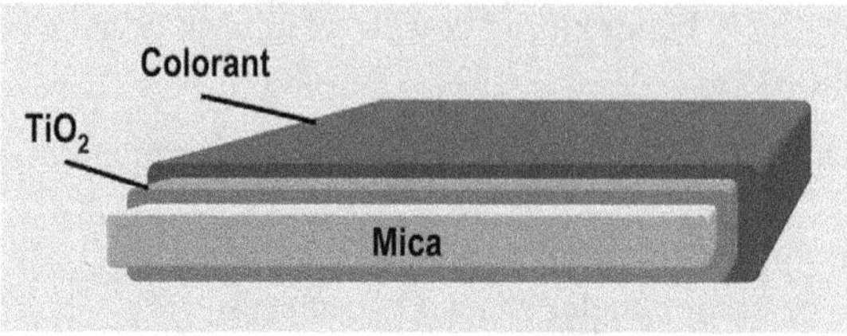

Figure 2.31: Schematic structure of a combination pigment

If the colors of the colorant and of the interference pigment are chosen to fit in a suitable manner, a clear two-color effect can be achieved. An example of this is the combination of interference blue with carmine-red. At the angle of reflection, the interference blue can be seen, whereas the red absorption color of carmine dominates at all other angles [58, 59].

Combination possibilities

The most important metal oxide in combination with TiO_2 on mica is iron(III) oxide in the hematite modification. A wide range of brilliant gold shades result from the combination of titanium dioxide and iron oxide. The spectrum ranges from pale yellow-gold and brilliant red gold to greenish gold, which looks amazingly like real gold. The pigments vary in their chemical composition and manufacturing process. Figure 2.32 shows the structure of three different gold pigments.

The precipitation of iron(III) oxide onto a titanium dioxide mica pigment produces pale golden to red golden products, depending upon the TiO_2-layer thickness. Because Fe_2O_3 also has a high refractive index, the combined effects of TiO_2- and Fe_2O_3-layers create the resulting interference colors. The absorption pigment iron(III) oxide determines the body color.

The simultaneous precipitation of iron and titanium oxide hydrates onto mica leads to a highly refractive iron titanium oxide after calcination. The greenish gold color arises from high refractive index yellow Fe_2TiO_5 phases that have been identified [60].

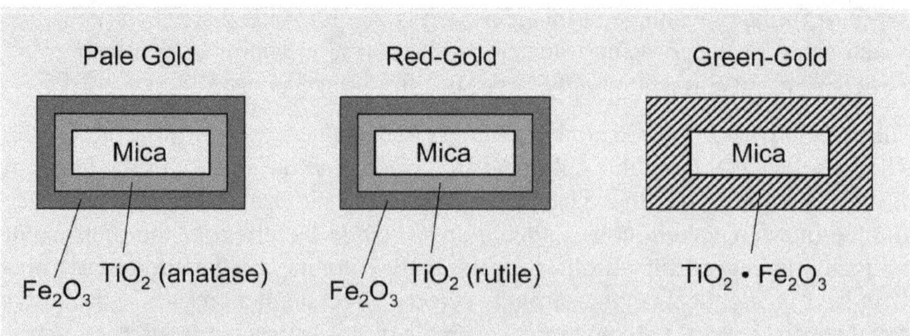

Figure 2.32: Schematic structure of three different gold pigments

Not only gold shades can be obtained from the combination of titanium dioxide and iron(III) oxide. Variations in layer thickness of both oxides give a wide spectrum of colors and interference effects ranging from yellow to red, red-brown, violet, blue and finally green. Theoretically, the range of color shades possible is virtually unlimited.

Other metal oxides – provided the refractive index is high enough – can also be used as a second color component for combination pigments. Blue-green to moss-green pigments are possible by the combination of titanium dioxide and chromium oxide.

Other inorganic pigments such as Prussian blue, cobalt blue, Fe_3O_4 and carbon black also produce impressive color effects in combination with titanium dioxide on the substrate. The latter two pigments are particularly interesting because the interference color is perceived as the absorption color.

The optical properties are similar to those of carbon black containing pigment mixtures in formulations, where the transmitted part of the light is absorbed [61]. The coating of TiO_2-mica pigments with organic colorants can take place using two methods:

1) Precipitation onto the TiO_2-surface in an aqueous suspension in the presence of complex surface-active substances.

2) Coating the colorant as a mechanically stable layer by using suitable additives together with physically mixing.

Combination pigments are frequently used in printing inks and cosmetic formulations. In cosmetics, red and two-tone colors are particularly popular. Due to regulatory restrictions, only a limited number of organic colorants such as carmine red or thioindigo can be used. Manufacturing processes are comparatively complex and must be carefully controlled to meet high purity and regulatory requirements.

2.8.4 Mica pigments with multilayers

New types of effect pigments which possess more than one metal oxide layer on the substrate platelet, have been developed in recent times, in addition to metal oxide mica pigments [18, 62]. This development is based on the recognition that the intensity of interference colors is strongly dependent on the difference between the refractive indices of neighboring optical layers and on the number of such layers. A larger number of layers with alternating refractive indices (preferred alternating high-refractive index and low-refractive index layers) increase the part of the light that is reflected from the pigment surface and therefore the intensity of the interference color.

A typical layer structure on mica is the sequence of a high-refractive TiO_2-layer, a low-refractive SiO_2-layer and of an additional high-refractive TiO_2-layer. Such a structure represents a three-layer optical system on both sides of the mica platelets.

Pigments with a more intense interference color or a more pronounced color travel can be manufactured when the thicknesses of the TiO_2- and of the SiO_2-layers are chosen in a suitable manner. In comparison with interference pigments with only one TiO_2-layer, multilayer mica pigments achieve a distinctly better intensity and brilliance of the interference color and excellent transparency. This is the result of the improved reflection profile of the individual pigment particles as shown in Figure 2.33.

The application properties of multilayer pigments are similar to those of mica pigments with single layers. Other substrates such as aluminum oxide, silicon dioxide, or borosilicate in the shape of thin platelets can be used in place of mica. Interesting effect pigments with intensive golden color shade are formed when using suitable combinations of iron oxide and silicon dioxide on the substrates mentioned before [63].

2.8.5 Pigments for outdoor applications

Special effect pigments suitable for outdoor applications must meet the highest standards for color fastness and weather resistance. Frequently they must be coated with an additional protective layer. For this purpose, special types of pigment have been developed and find particular application in automotive coatings. These products were tested comprehensively over many years by the automotive coatings

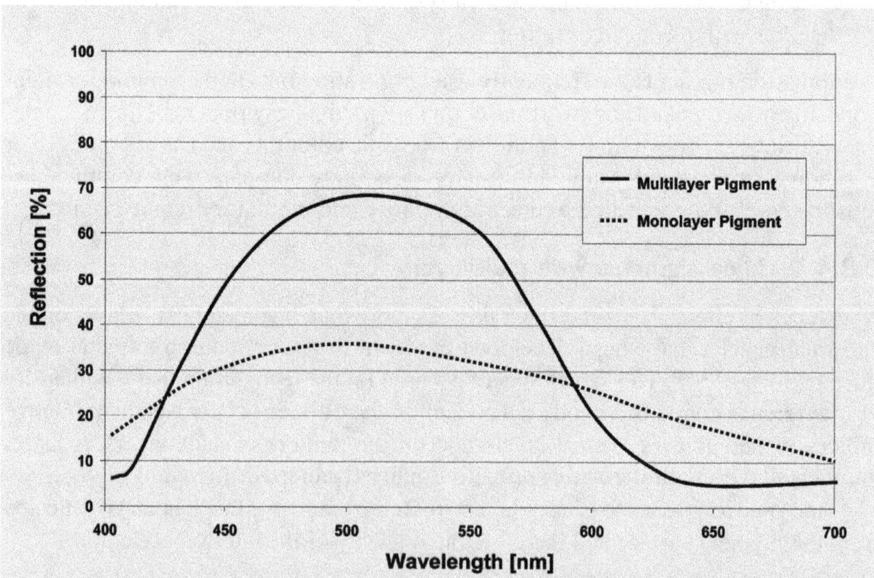

Figure 2.33: Calculated reflection spectra of one-layer and three-layer systems with reflection maxima in the green range of spectrum

producers and car manufacturers as well as by the pigment producers and can be considered as highly sophisticated [2, 21].

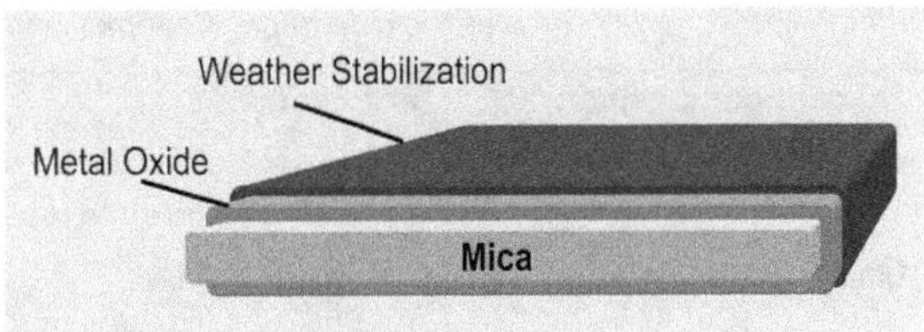

Figure 2.34: Schematic structure of a weather resistant metal oxide mica pigment

The metal oxide mica pigments are coated additionally with thin layers of transparent and colorless inorganic and/or organic compounds, e.g. silanes (Figure 2.34). These layers increase the light resistance by reducing the photo activity of the titanium dioxide surface. Additionally, this provides an optimized interaction between pigment and binder. This is also true for iron oxide or chromium oxide pigments [43, 64].

Finally, the pigments and the corresponding paints most go through the so-called "Florida test". In this case, test lacquer coatings are subjected to the sunny and humid climate of Southern Florida for at least two years and must not show any change in color or visibly deteriorate in any other way.

2.8.6 Mica pigments for new optical effects

2.8.6.1 Silver-gray and black pigments

Gray and black pigments are interesting color shades which enjoy great popularity, especially in industrial and automotive coatings. Same gray shades can be produced by mixing silver-white pearl luster pigments with carbon black [43].

These mixtures are very difficult to process, since the carbon black particles can agglomerate in the application system. This explains the search for alternatives to the silver-white pigment-carbon black mixtures in creating gray shades in the field of transparent effect pigments. Table 2.4 contains a list of mica-based effect pigments with special colors in the silver and gray area.

Manufacturing and properties

One possible route to manufacture is by the precipitation of ultra-fine carbon black particles onto a TiO_2-layer on mica. Even simpler is by a mixed precipitation reaction, which provides a uniform dispersion of carbon black particles in a TiO_2-layer (Figure 2.35) [70, 71]. If one compares such a carbon inclusion pigment with a mechanically processed mixture of carbon black and titanium dioxide, the carbon inclusion pigment demonstrates a greatly improved reflection at a glancing angle. However, both pigments demonstrate a similar spectral reflection curve for the down-flop condition (Figure 2.36).

Table 2.4: Examples for metal oxide mica pigments with lustrous silver and gray shades

Pigment composition	Preparation	Coloristic properties
TiO_{2-x}/TiO_2/mica [65, 66]	TiO_2/mica + H_2 (Ti, Si) T > 900 °C (solid state reaction)	gray, blue-gray
TiO_xN_y/TiO_2/mica [67]	TiO_2/mica + NH_3 T > 900 °C (solid state reaction)	gray, blue-gray
$FeTiO_3/TiO_2$/mica [68] (ilmenite-pigments)	Fe_2O_3/mica + H_2 T > 600 °C (solid state reaction)	silver-gray
Fe_3O_4/mica [67, 69]	Fe_2O_3/mica + H_2 T > 400 °C (solid state reaction) Fe^{2+} + O_2 + mica (wet chemical process) $Fe(CO)_5$ + O_2 + mica (CVD-process, fluidized bed)	black
TiO_2/C/mica (carbon inclusion pigments) [18, 70, 71]	$TiCl_4$ + C + mica (wet chemical process, calcination under N_2)	silver-gray

The smaller particle size (< 100 nm) of the precipitated carbon black may contribute to the higher luster of carbon inclusion pigments, since such small particles make no contribution at all to light scattering. With mechanically distributed carbon black particles in silver-white pigments, a different situation exists: The agglomeration of carbon black is practically unavoidable. The advantage of carbon inclusion is that the carbon black particles cannot separate from the pigment.

Another group of silver-gray pigments are the iron titanate mica pigments, frequently called ilmenite pigments. In this case, a mixed precipitation of titanium and iron oxide hydrate is carried out onto mica and then calcined in a reducing atmosphere of nitrogen and hydrogen. Silver-gray shining platelet pigments are formed. The reaction process can be described as [68]:

$$Fe_2O_3/TiO_2/\text{mica} + H_2 \xrightarrow{>600°C} FeTiO_3/TiO_2/\text{mica} + \ldots$$

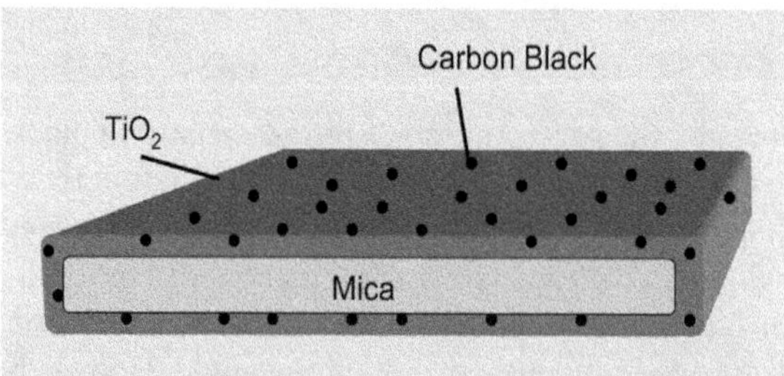

Figure 2.35: Schematic structure of a carbon inclusion pigment

Powder X-ray diffraction analysis of these pigments confirms characteristic reflections for mica and titanium dioxide as well as ilmenite. The pigments demonstrate a highly intense flop with changing viewing angle. This is characterized by the enhancement or dampening of the metallic-like gray shade, which is similar to the appearance of carbon inclusion pigments.

When titanium dioxide mica pigments are reduced in a solid-state redox reaction with reducing elements such as silicon powder under nitrogen, interesting blue-gray pearl luster pigments are formed [65]:

$$TiO_2/mica + Si \xrightarrow{>600°C} TiO_{2-x}/TiO_2/mica + ...$$

The blue-gray color results due to the development of titanium suboxides on the mica surface. Ti_2O_3 and γ-Ti_3O_5 are examples of such suboxide phases. The reaction can be accelerated by adding halides such as calcium chloride, allowing it to take place at a lower temperature. An intense color flop is also typical for titanium suboxide mica pigments [65]. Pigments with similar colors can be obtained using the following reactions [24, 25, 72]:

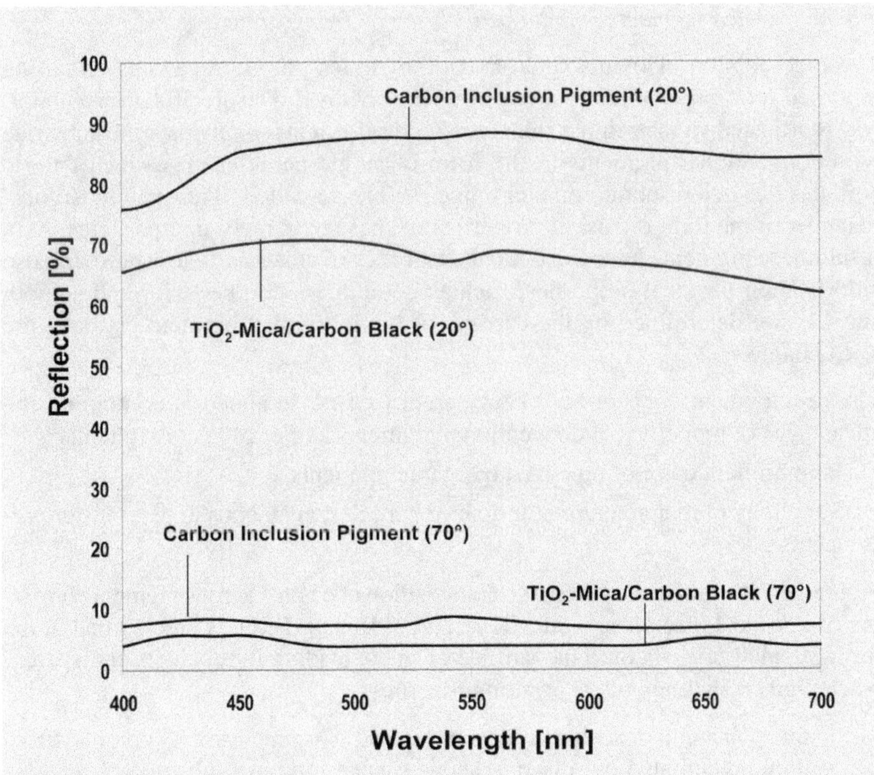

Figure 2.36: Reflection curves of a carbon inclusion pigment and a carbon black/titanium dioxide mica pigment using various illumination angles

$$TiO_2/mica + H_2 \xrightarrow{>900°C} TiO_{2-x}/TiO_2/mica + ...$$

$$TiO_2/mica + NH_3 \xrightarrow{>900°C} TiO_xN_y/TiO_2/mica + ...$$

The color of all of these pigments results from the combination of two effects: The absorption color of titanium suboxides or oxynitrides and the interference color, depending upon the thickness of the coating.

Black pearl luster pigments can be manufactured using Fe_3O_4-coatings on mica. Two processes can be followed:

1) Reaction of Fe_2O_3 coated mica with hydrogen at 400 °C [69]:

$$3\,Fe_2O_3/mica + H_2 \xrightarrow{400°C} 2\,Fe_3O_4/mica + H_2O$$

2) Application of Fe_3O_4 by precipitation from solutions of iron(II) salts [69, 73]:

$$3\,Fe^{2+} + ½\,O_2 + 9H_2O + mica \xrightarrow{pH8} Fe_3O_4/mica + 6H_3O^+$$

2.8.6.2 *Mica pigments without pearl luster: "Transparent Colors"*

The precipitation of oxides and/or oxide hydrates onto a mica surface can be modified such that non-lustrous pigments are obtained. The precipitation technology is adjusted so that, rather than creating uniform layers, it coats the substrate with conventional pigments in the form of single particles. The size of these particles lies in the submicrometer range [74]. The so-called "Transparent Colors" result from the light diffusion properties of these extremely fine particles. As a result these pigments have the optical properties of small particles (pure, intense color). Their physical properties, such as good dispersion behavior and smooth surface, are determined by the carriers of the individual pigment crystals, the mica platelets.

The development concept for "Transparent Colors" is aimed at combining the advantages of properties of conventional pigments and pearl luster pigments:

- Pure brilliant color of nm-sized inorganic pigments,
- Simplicity of manufacture of µm-sized particles, analogous to pearl luster pigments.

Figure 2.37 illustrates the schematic composition of a non-lustrous pigment consisting of mica and black iron oxide (Transparent Black). Fe_3O_4 is precipitated in the form of small crystals onto the surface of mica, leading to relatively transparent black pigments with good dispersion properties.

The main application of such pigments is the field of cosmetics. Any color additive that can be precipitated out of an aqueous suspension onto substrate platelets is suitable. Some examples include iron oxide (red, yellow and black), Prussian blue, chromium oxide and titanium dioxide [59, 75-77]. The resulting color effect depends upon the amount, particle size, particle shape and chemical composition of the precipitated inorganic substances.

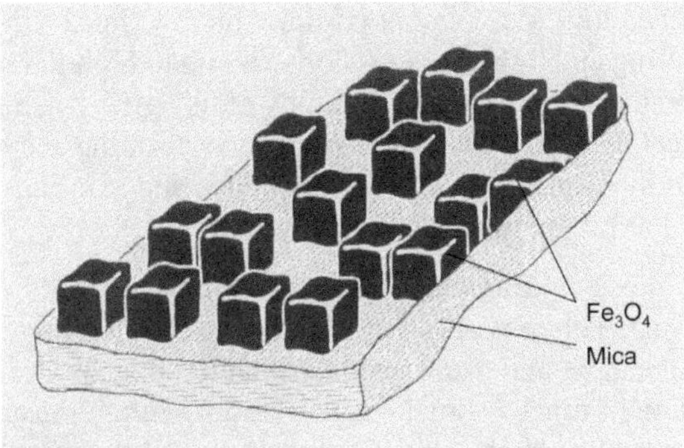

Figure 2.37: Schematic illustration of a non-lustrous magnetite mica pigment

Compared to conventional pigments, "Transparent Colors" have the advantage of good dispersion without milling. With conventional pigments the color shade varies greatly with dispersion and is therefore very difficult to control. When using standard milled pigments, it is almost impossible to achieve an optimum degree of extinction. In the case of "Transparent Colors", on the other hand, a completely reproducible color development can be obtained due to a controlled precipitation of the oxide crystals. Figure 2.38 shows the electron micrograph of a transparent red pigment where single Fe_2O_3-crystals are coated onto the surface of mica.

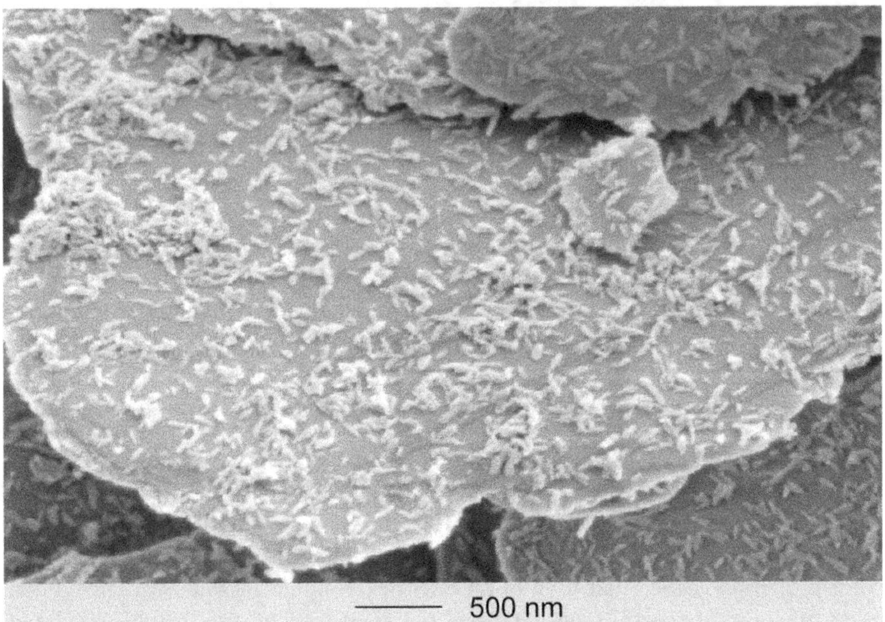

Figure 2.38: Scanning electron micrograph of a "Transparent Color" pigment

The ultra-fine pigment particles and the mica substrate together create light scattering and light transmission, leading to highly transparent pigments. Used in cosmetic products, “Transparent Colors” give a natural coloring and a pleasant skin feeling. The light scattering properties of these pigments smooth out the variation in the skin colors and help wrinkles and skin imperfections to visually disappear. The skin maintains a natural look.

For cosmetic applications, a lower mica particle size distribution (< 20 μm) is often used. This not only permits a good skin feeling, but also helps towards smooth spreading and a uniform distribution upon application. Due to the good skin adhesion of mica, cosmetic formulations with “Transparent Colors” have an excellent durability on the skin. “Transparent Colors” are always used in cosmetic applications when little or no luster is desired. They can be incorporated in combination with effect pigments or conventional pigments. Suitable applications are loose or pressed powders, rouge, lipsticks or emulsions.

2.8.6.3 *Mica pigments with low pearl luster: “Low Luster Pigments”*

The deposition of inorganic compounds upon mica substrates also develops pigments, which scatter light more than they reflect it. Luster pigments based on titanium dioxide/mica can amplify skin imperfections when applied in cosmetic formulations. This is particularly obvious with wrinkles, where light is reflected at certain angles and the edges of the wrinkles are emphasized.

Attempts have been made to smooth out these unwanted imperfections with highly refractive conventional pigments. This type of pigmentation leads, however, to an unnatural, “mask” appearance. A compromise between these two extremes was found with the use of transmitting, light scattering pigments, which lessen the visibility of wrinkles [59].

In “Low Luster Pigments”, barium sulphate and titanium dioxide are precipitated together onto very fine particle size mica to create a light scattering pigment. When preparations controlling these pigments are applied, the skin remains natural looking [78, 79]. The platelet structure of the mica gives the pigments a pleasant feel to the skin. During application, the particles align well and adhere to the skin. An additional advantage, as in all mica pigments, is the good dispersibility in the application media.

2.8.7 Functional metal oxide mica pigments

Metal oxide mica pigments were originally developed because of their excellent color properties. Since that time in product development, it was recognized that other functional properties can be easily obtained through use of the same layer substrate technology.

Table 2.5: Functional pigments based on the layer-substrate-principle

Functional property	Composition	Application fields
electrical conductivity	$(Sn,Sb)O_2$/mica [70, 81 - 84]	antistatic flooring, electrostatic painting
magnetism	Fe_3O_4/mica [43]	magnetic surfaces
solar heat reflection	TiO_2/mica [80, 85, 86]	greenhouses, transparent roof lights
laser marking of polymers	TiO_2/mica $(Sn,Sb)O_2$/mica [18, 80, 87]	laser marking of plastics
mechanical reinforcement	TiO_2/mica [88, 89]	filler for improvement of mechanical properties of polymers

Pigments are basically chemical substances consisting of particles, which are practically insoluble in application media and which are used as colorants or due to their corrosion-inhibiting, magnetic, electrical, or electromagnetic properties [1]. This definition confirms, that pigments are not only of interest for their color, but also because of various physical characteristics, which can not be derived from the original term "pigmentum" (Latin: painter's color). Such materials can usefully be described with the expression "functional pigments" [80].

New pigments have been developed in the last years and introduced to the market along with well-established conventional functional pigments. These new materials are mostly based on natural mica platelets as substrate. They contain metal oxide layers on the mica, which possess electrically conductive, magnetic, infrared reflecting or other physical properties and give rise to the functional behavior of the pigments. Table 2.5 contains a summary of important functional pigments based on mica.

Some of the functional pigments based on mica are described in the following chapters.

2.8.7.1 Light colored, electrically conductive pigments

Until recently, there were hardly any electrically conductive, light-colored alternatives to carbon black, but now the development of the layer-substrate principle has made such pigments available [70, 81–84]. The inert support is made of mica. This is coated with a thick layer of metal oxides, the conductive component of which is mainly tin oxide containing antimony. Whilst SnO_2 has a high resistance, compositions with the general formula $(Sn,Sb)O_2$ show semi-conductive to conductive properties. Antimony is built into the SnO_2-lattice in positions taken by tin.

Light-colored, conductive pigments of this form are available under the trade name "Minatec" [90]. There are light gray, platelet-shaped pigments in the "Minatec" series.

They are appropriate for coatings with intense colors and formulae that require a high level of transparency (Figure 2.39, "Minatec" 31 CM).

Special pigment types can be used mainly for conductive coatings that demand higher hiding power. By introducing an extra layer of TiO_2 between the mica and the $(Sn,Sb)O_2$, part of the light is reflected at 450 nm (Figure 2.40, "Minatec" 30 CM).

In contrast to the two products mentioned above, "Minatec" 40 CM is not composed of 100 % platelet-shaped particles. A proportion of them are spherical. "Minatec" 40 CM is particularly suitable for thick, conductive coatings for example for flooring.

Conductive pigments made in this way are color neutral, which gives them a decisive advantage over carbon black. This enables light colored conductive fillers and coatings with antistatic and electrostatic-dissipative properties (ESD) to be formulated. "Minatec" pigments can be used in industrial applications in a wide resistance range from 10^4 to 10^9 Ω. This includes both uses for electrical conductivity and antistatics. The pigments demonstrate outstanding formulating properties, particularly in two areas:

- as an additive for conductive fillers for use in ESTA application.
- as the conductive component in coatings with permanent antistatic or ESD properties (flooring, explosion protection, clean rooms) see Figure 2.41.

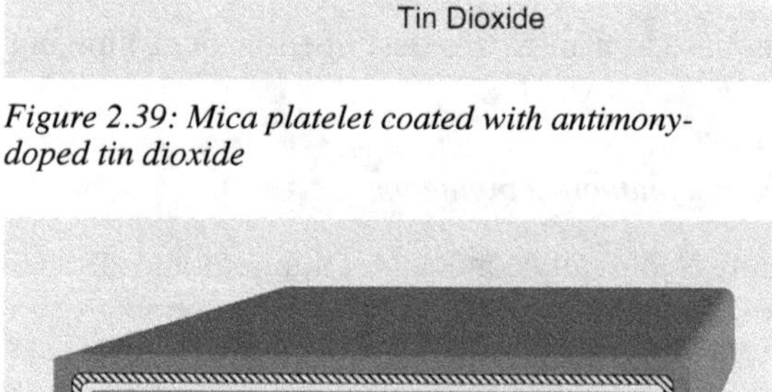

Figure 2.39: Mica platelet coated with antimony-doped tin dioxide

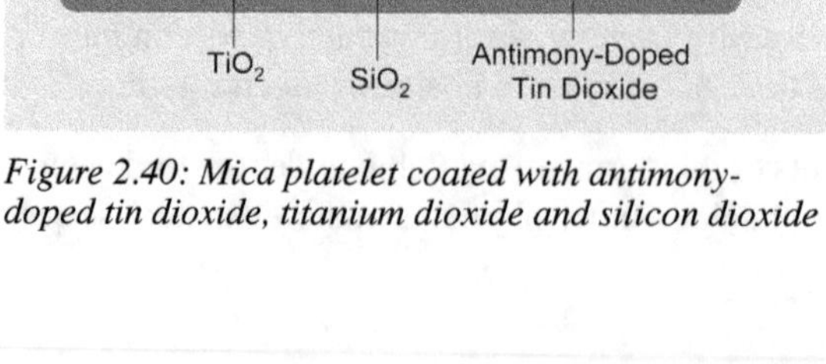

Figure 2.40: Mica platelet coated with antimony-doped tin dioxide, titanium dioxide and silicon dioxide

The use of a conductive filler improves the efficiency of the basecoat application by significantly reducing overspray.

In order to be conductive, the pigment particles must touch one another as much as possible. Thus a certain minimum concentration is needed to create a conductive path for the electron transport. The amount of pigment needed depends on the particle shape, its distribution in the medium and the latter's composition.

Conductive pigments are produced in the following way: Solutions of the salts of

Table 2.6: Characteristics of the light-colored conductive pigments "Minatec" 30 CM and "Minatec" 31 CM

	"Minatec" 30 CM	"Minatec" 31 CM
% by weight mica	41	59
% by weight $(Sn,Sb)O_2$	43	41
% by weight TiO_2	12	-
% by weight SiO_2	4	-
ratio Sn:Sb	85:15	85:15
particle diameter	<15 µm	<15 µm
color	light gray	transparent/light gray
powder resistance	<100 Ω * cm	<140 Ω * cm

either tin and antimony and silicon and titanium are slowly added at the same time as a base to a water-based mica suspension. Thus the oxide hydrate of the metals is deposited on the finely dispersed mica particles. The layer, which adheres well to the mica particles, is not yet conductive. This property is created after the subsequent calcination process which converts the oxide hydrate into an oxide by dehydration and the antimony is fixed in the SnO_2-lattice.

The properties of "Minatec" 30 CM and "Minatec" 31 CM are summarized in Table 2.6. The powder resistance of these pigments is as follows [43]:

- As the thickness of the $(Sn,Sb)O_2$-layer increases, and the ratio of tin to antimony remains constant, the conductivity rises. Above a layer thickness of 20 nm, there is no further improvement in electrical conductivity.
- If the Sn:Sb ratio is changed to give more antimony, the conductivity increases. The optimum conductivity value is Sn:Sb = 75:25. The increasing level of antimony not only manifests itself in a reduction in powder resistance, the surface conductivity also increases, a property which is interesting for the user.
- If coarser grades of mica are used, an acceptable resistance is achieved at a comparably lower level of antimony. An increase in the antimony level brings no further notice-

Figure 2.41: Floor covering in a high rack warehouse antistatically equipped with a light colored, electrically conductive pigment (surface resistance 10^4 Ω)

able reduction in resistance. When fine mica grades are used, more metal oxide is needed to guarantee complete coverage of the particles, because of the larger, specific surface area. However, the conductivity is better.

Similar results are achieved if kaolin or other platelet-shaped substrates are used instead of mica [83].

Electrically conductive metal-oxide mica pigments exhibit a number of special features during use when compared to substrate-free $(Sn,Sb)O_2$-powders. The latter often tend to form agglomerates and are difficult to disperse. This is, however, not the case with mica pigments, as good dispersion can be achieved merely by stirring them into the paint system.

2.8.7.2 *Magnetic pigments*

Magnetic pigments can also be produced using the layer-substrate principle [43]. Examples of conventional magnetic pigments are: γ-Fe_2O_3, Co-containing γ-Fe_2O_3, Fe_3O_4, Co-containing Fe_3O_4 and CrO_2. Iron pigments are also used as magnetic powder. They all have needle-shaped particles with an average length of 200 to 600 nm. They are mainly used for magnetic-recording mediums. Barium-ferrite pigments with their hexagonal structure are also becoming increasingly important for this use. The preferred diameter of ferrite platelets is between 40 and 300 nm.

Magnetite is particularly suitable for use as the coating for magnetic pigments with mica substrates. These materials can be manufactured either by the reduction of Fe_2O_3 containing mica pigments as described in Chapter 2.8.6.1 or by precipitation and oxidation starting from iron(II) salt solutions. Typical particle sizes of such pigments are 10 to 60 µm. Magnetite gives these pigments their black absorption color; however at certain film thicknesses they appear colored at the glancing angle. Further properties of these pigments are a matt luster effect and good hiding power. At present no uses for magnetic pigments of the layer-substrate type have been found that are comparable to those for conventional materials.

2.8.7.3 *Pigments for the laser marking of plastics*

"Laser-markable pigments" are also based on the layer-substrate principle that is used in special effect pigments [70, 80, 87]. In this case, thin mica platelets are coated with a specially prepared layer of either TiO_2 or $(Sn,Sb)O_2$ ("Lazerflair"-pigments) [91]. The principle of how laser-sensitive pigments, the marking process and laser welding work is described in Chapter 4.4.

2.8.7.4 *Solar heat-reflecting pigments*

"Solarflair" 870 and "Solarflair" 875 are translucent heat-reflecting pigments that are based on metal-oxide coated mica [92]. Whilst they reflect infrared radiation, they are almost completely transparent in the visual range. They are highly suitable for architectural and horticultural use, where it is important to have plenty light, but little heat. Whether in light domes or greenhouses, for shutters or conservatories, "Solarflair" pigments absorb only a small amount of light and substantially reduce the heating caused by intense sunlight. The conditions, under which they are used, can make a difference of up to 10 °C when compared to unpigmented systems [86, 93].

Figure 2.42: Pedestrian bridge near Lyon (France) pigmented with "Solarflair" 875

Figure 2.42 shows an example of the use of "Solarflair" 875. This pigment is particularly suitable for use in building and living areas, because the maximum transmission wavelength area is in the green range where the eye has the highest light perception (Figure 2.43).

"Solarflair" 870 is most appropriate for greenhouses. Its maximum transmission wavelength area is in the optimum range for photosynthesis (see Figure 2.44, page 72). In principle "Solarflair" 870 is also suitable for the building and living areas, but the heat reflection is somewhat less than "Solarflair" 875; however the pigment is more color neutral.

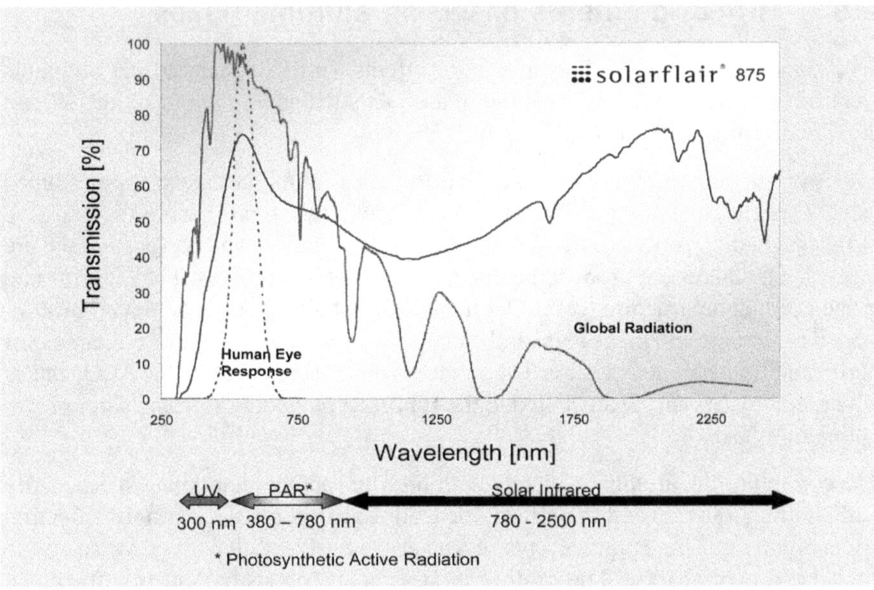

Figure 2.43: Transmission spectrum of the heat-reflecting pigment "Solarflair" 875

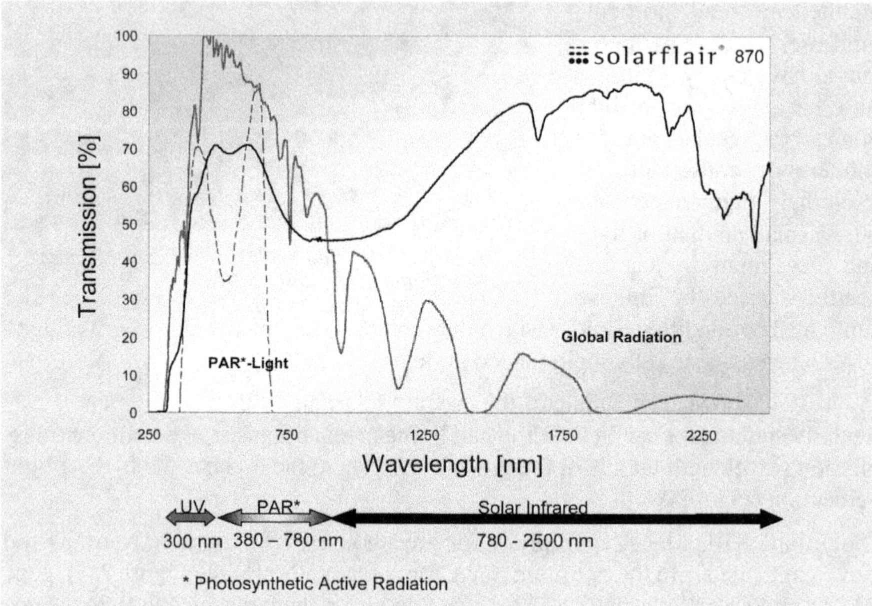

Figure 2.44: Transmission spectrum of the heat-reflecting pigment "Solarflair" 870

2.9 Effect pigments based on alumina flakes

To expand the range of achievable color effects, further special effect pigments were developed, which use substrate materials such as aluminum oxide, silicon dioxide or borosilicate instead of mica [21, 94].

Effect pigments based on aluminum oxide flakes (alumina flakes) are produced using a special, technologically ambitious manufacturing process. In the first step of the production process, Al_2O_3-flakes with a very smooth and perfect surface are formed. The flakes are produced using a crystal growth process. They are formed in the corundum structure (α-Al_2O_3). The desired flake shape is achieved by controlled crystal growth in the axial and equatorial directions. The extraordinary color purity and transparency of the effect pigments obtained by coating the Al_2O_3-flakes with metal oxides can be attributed to the synthesis procedure yielding single-crystalline thin flakes [95–98].

The coating of the alumina platelets with high-refractive metal oxides, especially with titanium dioxide and iron(III) oxide leads to extraordinary strongly reflecting effect pigments. These pigments possess a very intensive glitter effect, the so-called crystal effect or sparkle. The coating process is analogous to that used for metal oxide mica pigments except it starts from an aqueous suspension of Al_2O_3-flakes.

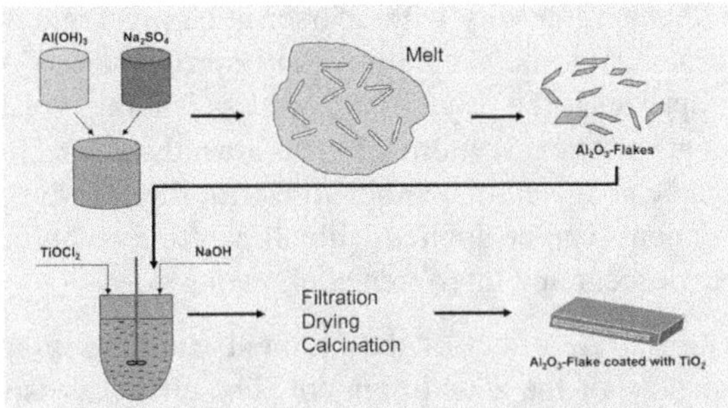

Figure 2.45: Process scheme for Al_2O_3-flake pigments

The complete manufacturing process of pigments based on aluminum oxide platelets, consisting of the flake production and the coating with metal oxides is shown in Figure 2.45 [96].

The effect pigments so described are available on the market as "Xirallic" pigments (High Chroma Crystal Effect Pigments) [99]. They are modified for outdoor applications with an additional weather stabilizing surface treatment. Figure 2.46 shows electron micrographs of pure and of TiO_2-coated Al_2O_3-flakes.

Silverwhite and interference pigments based on Al_2O_3-flakes can be produced – analogous to similar pigments based on mica platelets – in different colors, e.g. gold, red, blue or green, by variation of the titanium dioxide layer thickness. Depending of the layer thickness, the coating with iron oxide leads to bronze-, copper- or red-colored effect pigments. The pigments obtainable at the market possess a narrow particle size distribution of about 5 to 30 µm as well as a high aspect ratio. The sparkle effect referred to above has its origin in the optimized thickness of all layers in the pigment structure, including the Al_2O_3-substrate. Analytical investigations show that the thickness range of the aluminum oxide platelets and the resultant pigments can be controlled very precisely [22, 110].

The luster effects of the Al_2O_3 based pigments are additionally intensified by directed illumination, e.g. by sunlight. Previously, the resultant sparkle effect could not be achieved with small-sized effect pigments. This is especially important for automotive coatings, because it was not possible before to obtain a sparkle effect

Figure 2.46: Scanning electron micrographs of pure (left) and TiO_2-coated (right) Al_2O_3-flakes

with such intensity with coarser mica pigments without a substantial loss of performance at the same time. An improvement of the application and of the color properties is the result of the process leading typically to minimized superfine pigment particles. Scattering effects are reduced and intensive, pure interference colors can be achieved in combination with the colorless aluminum oxide substrate. The pigments can be applied with all painting techniques established for conventional pearlescent and interference pigments.

The sparkle effect of the pigments in coating applications is controlled by the quantity of the added pigment. The effect is visible already at concentrations of 0.1 % in the paint system. The intensity is steadily increased up to pigment concentrations of about 2 %. The single light spots arising from the pigment structure and orientation in the paint system seem to spring back and forth when a painted metal plate is tilted. Also the color of the other components of the paint has an influence in addition to the concentration of the pigment. A dark color, for example, forms a high contrast to the bright light spots and strengthens the effect. Favorable is also the use of a covering varnish in order to achieve an optimal effect [22, 100].

Effect pigments based on alumina flakes may be applied in all substantial systems, i.e. in coatings, plastics, printing inks and cosmetic formulations. An additional surface treatment is applied to the pigment particles to adjust them to solvent-based and water-based coating systems. The pigments can be used in combination with many conventional pigments as well as with other effect pigments. Nearly unlimited styling possibilities are the result of these combinations [22, 100].

2.10 Effect pigments based on borosilicate flakes

The most recent development in the field of transparent effect pigments has led to pigments based on borosilicate glass. Thin glass platelets are completely colorless and transparent. These favorable properties can be used for the manufacture and application of special effect pigments. Coating of the glass flakes with TiO_2 leads to very pure interference colors without any mass tone color [21, 22, 101].

The manufacture of borosilicate pigments is carried out in three steps:

- production of the raw flakes,
- grinding and classification of the flakes,
- coating of the flakes with highly-refractive metal oxides (titanium dioxide, iron(III) oxide).

The synthesis of the borosilicate platelets is carried out starting from a melt. The purity of the raw materials used guarantees that the resultant flakes achieve the desired colorlessness and transparency. The flakes are produced directly from the melt. They can be used after cooling-down, grinding and classification as substrate platelets. Coating of the borosilicate flakes is analogous to the processes used for mica and aluminum oxide substrates:

- coating of the flakes with metal oxide hydrates in aqueous suspension,
- separation of the coated flakes by filtration,
- calcining of the particles to transform the metal oxide hydrates into highly-refractive metal oxides.

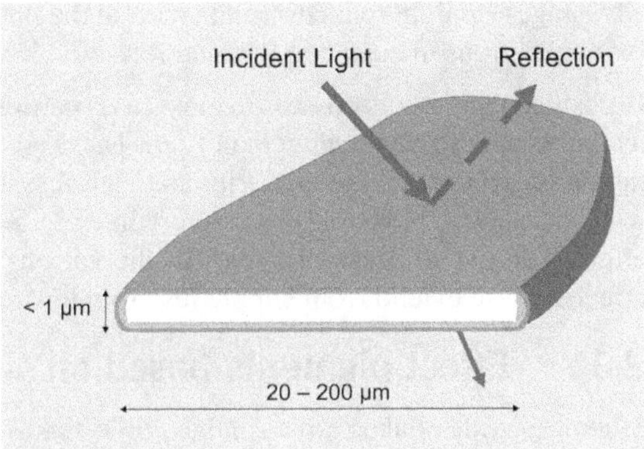

Figure 2.47: Schematic structure of a pigment particle based on borosilicate. Layer structure: TiO_2-SiO_2-borosilicate-SiO_2-TiO_2.

Figure 2.47 shows a schematic view of the structure of a borosilicate pigment from the product range "Miraval" [102]. This pigment consists of the borosilicate flake, a first layer of silicon dioxide and on top of this a second layer of highly-refractive titanium dioxide. The SiO_2-layer encapsulates the substrate and acts as a protective layer with an ideal surface for the coating with the TiO_2-layer, which determines the color. In addition, the SiO_2-intermediate layer provides the thin substrate platelets with mechanical stability. Furthermore, at the end of the process it leads to interference pigments with an improved luster. The reason for this is the low refractive index of the SiO_2 (1.42 to 1.47) in comparison to borosilicate (1.50 to 1.58). Therefore a larger difference between the refractive indices of the TiO_2-layer and the adjacent layer is achieved at the same time.

The advantageous coating is also visible in the electron micrographs in Figure 2.48. The edges of the pigment particles are perfectly covered and rounded. This has

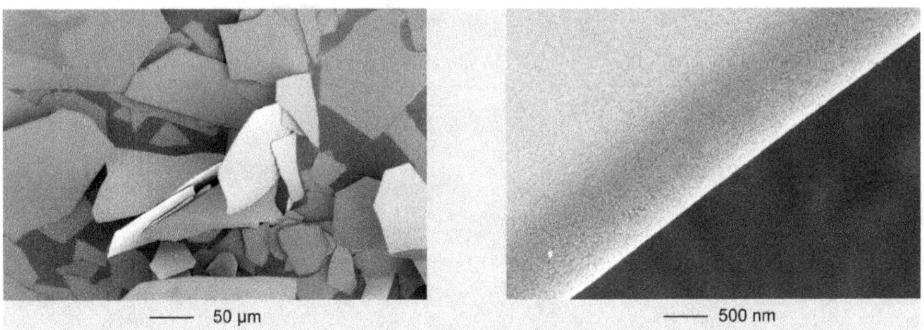

Figure 2.48: Scanning electron micrographs of pure (left) and TiO_2-coated (right) borosilicate flakes

advantages for light reflection and yields at the same time an improved skin feeling for applications in cosmetic formulations [101].

Borosilicate pigments possess in most cases particle diameters below 200 µm and a thickness with values smaller than 1 µm. These pigments provide a unique and striking multicolored sparkle effect in application systems. This optical phenomenon, which is especially visible with silverwhite pigments, is also called the multicolor effect. The use of these pigments in the various systems allows manifold color effects. These extend from simple luster up to strongly reflecting formulations.

2.11 Effect pigments based on silicon dioxide flakes

Silicon dioxides flakes (silica flakes, SiO_2-flakes) alongside the alumina and the borosilicate flakes, belong to the innovative substrate materials, which lead to extraordinary color effects in the field of transparent and semi-transparent effect pigments. The manufacture of the silica platelets acting as substrate is possible using a web coating process. This method allows the manufacture of thin flakes with completely uniform thickness. Silicon compounds, such as sodium waterglass, are used as precursors for the process. Figure 2.49 shows the scheme for the manufacture of the SiO_2-platelets as well as the successive step, where the flakes are coated with highly refractive metal oxides, especially with titanium dioxide and iron oxide [94, 96, 103, 104]. The structure of the pure SiO_2-flakes and the TiO_2 coated SiO_2-flakes can be seen in the electron micrographs in Figure 2.50.

Pigments based on silicon dioxide platelets are especially striking due to their multi-color effects (angle-dependent color, color travel effect, color flop effect). The precise adjustment of the thickness of the substrate and metal oxide coating

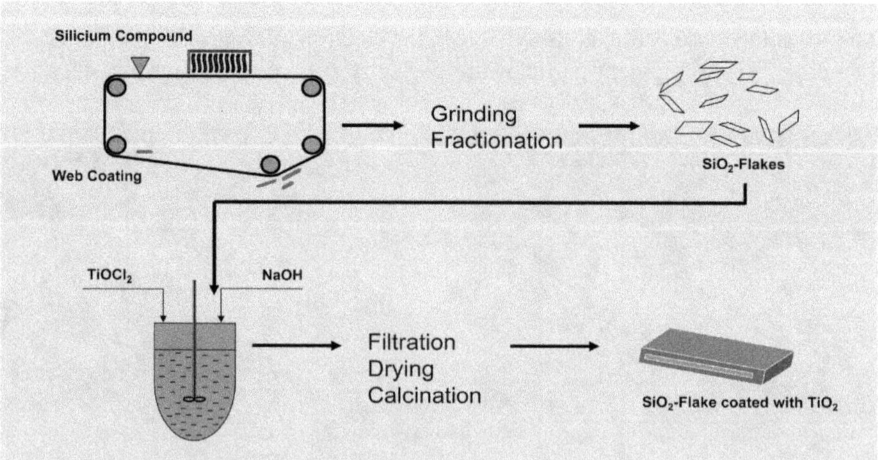

Figure 2.49: Process scheme for SiO_2-flake pigments

allows optimal development of the color properties of the pigments. The gradual color travel of the pigments enables at the same time pronounced multi-color changes when used in application systems. When used in combination with transparent colored pigments and black pigments extensive color effects arise, which offer color designers broad scope for creative styling.

Synthetically produced SiO_2-flakes demonstrate three main differences in comparison with mica platelets:

1. The thickness of the SiO_2-substrate can be controlled very precisely during manufacture. This is the reason why the final pigments are real optical three-layer systems. The interference color of these pigments is stronger than that of conventional mica pigments because mica possesses a broad thickness distribution as a result of its manufacturing process.
2. As synthetic substrates, SiO_2-flakes are free of iron impurities, which in the case of natural mica lead to a slightly yellow mass tone color.
3. SiO_2 has a lower refractive index (1.46) than mica (approx. 1.58) and leads therefore to more intensive interference effects [96]. Figure 2.51 shows a schematic view of a section through a mica platelet coated with a highly refractive metal oxide in comparison to a metal oxide coated SiO_2-platelet. The properties that can be achieved with SiO_2-flakes can be summarized as follows:

- improved substrate characteristics,
  - uniform and controllable substrate thickness,
  - smooth and uniform substrate surface,
  - transparent substrate with no mass tone color,
- improved reflection and refraction of light,
- improved chromatic strength and purity,
- new color travel properties.

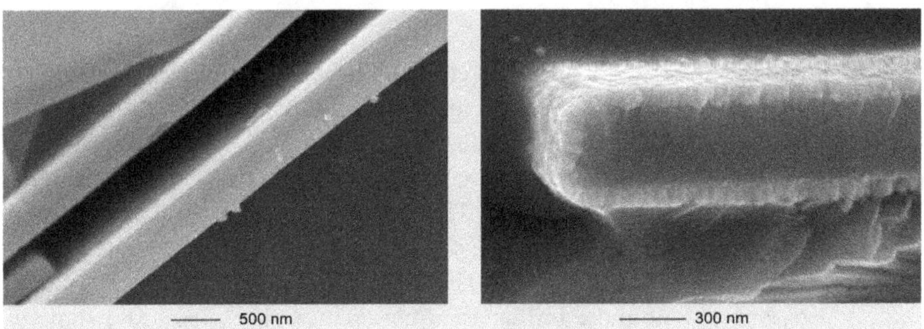

Figure 2.50: Scanning electron micrographs of pure (left) and TiO_2-coated (right) SiO_2-flakes

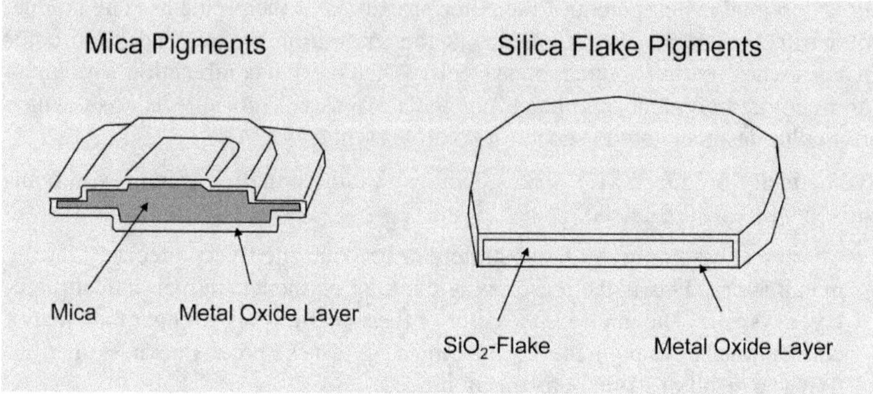

Figure 2.51: Schematic view of a section through a mica platelet coated with a highly refractive metal oxide in comparison to a metal oxide coated SiO_2-platelet

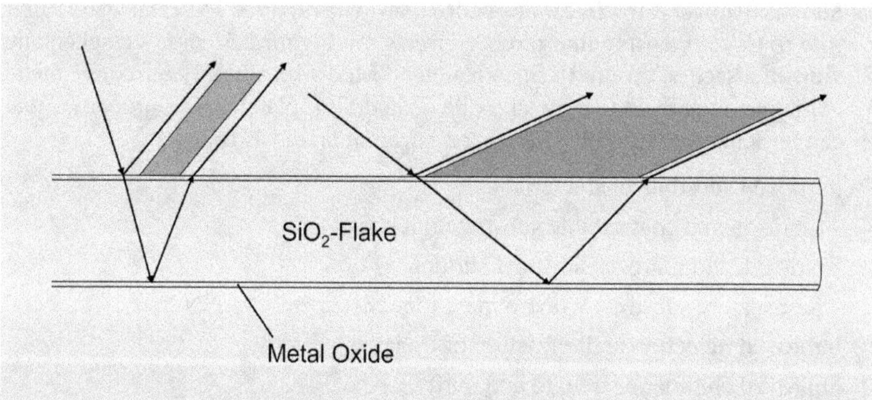

Figure 2.52: Color variation vs. viewing angle of a SiO_2-flake pigment with a very thin metal oxide layer (TiO_2, Fe_2O_3)

The thickness of the SiO_2-flakes produced using the web coating process can be selected in the range 50 and 1000 nm. The flakes act as an excellent substrate for coating with highly refractive metal oxides. The deposition of oxides like TiO_2 and α-Fe_2O_3 is given with the wet-chemical coating process established for mica pigments. Analogous to the mica substrates, TiO_2 is formed in the anatase modification working with direct coating. To generate rutile layers, a SnO_2-intermediate layer is necessary before the deposition of the TiO_2-layer. The SiO_2-flake pigments with rutile layers demonstrate more brilliant and chromatic color effects and show in addition higher light stability. Therefore most silica based pigments are produced with rutile layers.

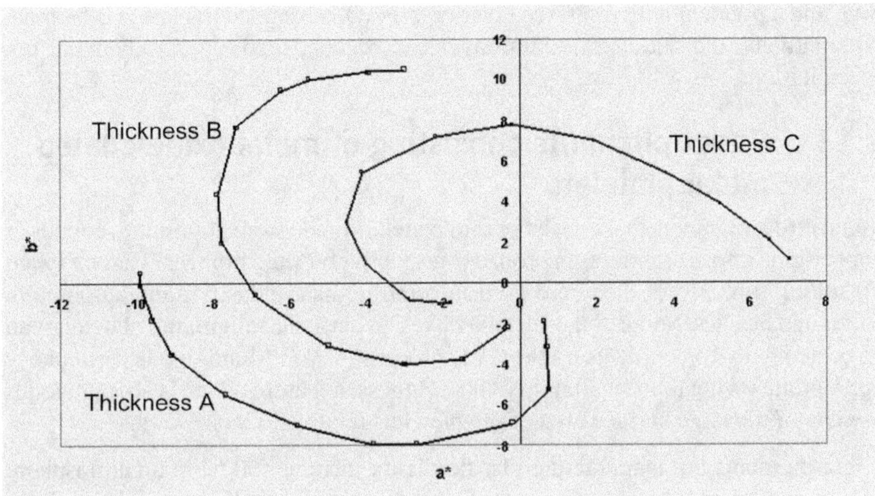

Figure 2.53: Angle-dependent color travel of Fe_2O_3-coated silica flakes of different thickness

The color travel effects described above can be adjusted across a broad color range. Color changes from gold-silver to green and further from green to blue or even dark blue are possible. Figure 2.52 demonstrates how the color flop is generated as a function of the viewing angle. The angle-dependent color travel in the CIELAB system of three pigments with constant Fe_2O_3-layer thickness on SiO_2-flakes of different thickness is shown in Figure 2.53.

2.12 Effect pigments based on iron oxide flakes

The iron oxide platelets described in Chapter 2.5 can be used for the manufacture of luster pigments with a stronger interference color and particular angle-dependent color travel. This is achieved by applying additional metal oxide layers on the iron oxide platelets [105]. Typically a layer structure of $Fe_2O_3/SiO_2/Fe_2O_3/SiO_2/Fe_2O_3$ is generated, whereby the central Fe_2O_3 layer (core) consists of the iron oxide flake, while the two outer layers provide selective reflection and together with the SiO_2-layers provide a strong color change in the color space violet-red-orange-gold. In addition this gives the pigments good corrosion stability.

The manufacture of such five-layer effect pigments takes place in a stirred vessel, where iron oxide platelets are coated with a precisely controlled SiO_2-layer by hydrolysis of tetraethyl orthosilicate in alcoholic suspension. After washing and drying, the outer Fe_2O_3-layer is deposited in a fluidized bed reactor using a CVD-process (Chemical Vapor Deposition). In this process gaseous $Fe(CO)_5$ is reacted with oxygen at raised temperature in the presence of iron oxide flakes coated with

SiO_2 and a pigment with a surface layer of α-Fe_2O_3 of defined thickness is formed. Alternatively, the outer iron oxide layer can be deposited by wet-chemical co-precipitation.

2.13 Effect pigments consisting of metal oxide-coated metal platelets

Metal effect pigments that consist of thin platelets made from aluminum, copper or copper-zinc alloys ("aluminum bronzes" respectively "gold bronzes") have existed for a long time [106–108]. They are used in printing inks, industrial and automotive paints and plastics. Some of these metal flakes, in particular aluminum platelets, can be coated with iron oxide using the CVD process [23, 67, 109]. Aluminum is particularly appropriate owing to its availability and lightness, a feature which is advantageous in view of the large surface area of the platelets relative to their weight.

These pigments are manufactured by fluidizing them at 450 °C in an atmosphere of nitrogen in a fluidized bed reactor. The reactor is essentially a cylindrical fluidized bed. The temperature is regulated through the walls. The reactants – $Fe(CO)_5$ and O_2 – are streamed in the reactor together with the agitated Al-flakes. For a high-quality coating, the reactants must be well and evenly dispersed in nitrogen. In some cases, steam is also added to reduce the build up of an electrostatic charge. The thickness of the deposited pigments is regulated by the reaction time in keeping with the equation:

$$2\ Fe(CO)_5 + 6.5\ O_2 + Al_{(pigment)} \rightarrow Fe_2O_3/Al_{(pigment)} + 10\ CO_2$$

The iron oxide is deposited on the aluminum, as hematite (α-Fe_2O_3). The resulting pigments display golden, orange and red effects, which are similar in appearance to metal. Table 2.7 shows the effect the film thickness of iron oxide on Al-platelets has on the pigment color [110].

Table 2.7: Color shade dependence of an Al-based pigment from the layer thickness of the deposited iron oxide [110]

1st order color	Layer thickness Fe_2O_3 [nm]	2nd order color	Layer thickness Fe_2O_3 [nm]
silver	0	gray-yellow	63
light yellow	10	gray-gold	75
gold	21	reddish gold	88
red-gold	28	red-gold	104
red	35	red	122
violet	44		
gray-violet	53		

The pigments in the gold and red-gold ranges give the highest color brilliance, because this is where the adsorption of the iron-oxide layer best matches the interference reflection [110].

If the aluminum flakes are coated with metal oxide or metal oxide hydrate before the CVD-process, the reactivity of the metal surface is reduced. Furthermore, it is possible to manufacture multi-layer pigments by using layers of SiO_2, that produce angle-dependent color effects (color travel pigments) [110].

Hydrogen is produced when aluminum platelets are coated in an aqueous suspension. It is therefore recommended that, beforehand, they are given at least a thin coating of metal oxide hydrate or metal oxide in an organic solvent. SiO_2-layers give interesting effects. If aluminum platelets are given a thick layer of SiO_2, followed by a metal oxide such as α-Fe_2O_3 with a high refractive index, the mirror-like reflection of the metal surface is superimposed by the absorption and interference caused by the interaction of the light with the metal oxides.

The wet-chemical coating of aluminum with SiO_2-platelets mostly takes place using an alcoholic suspension. If an aqueous suspension is to be used, the flakes need to be passivated beforehand. There is a noticeable difference between the optical effect of a thick SiO_2-layer and TiO_2. The low refractive index of SiO_2 causes a small degree of diffraction of light. Therefore the length of the light rays in the SiO_2-layer that land vertically or at an angle are different and cause changing interference effects. This creates angle-dependent color effects (compare Chapter 2.11). A different color can be seen from each angle.

The color of light reflected by a layer of SiO_2 depends on the angle from which it is seen, but its intensity is very small. If the interface of the SiO_2-layer is directly with the air, only 4 % of the light is reflected.

Thus the effect on the human eye is small. If, however the SiO_2-layer is covered with a thin layer of a material that has either a high absorption level or a high refractive index, the reflectivity of the SiO_2-layer can be greatly increased (*Fabry-Perot* principle). In general this reflection promoting layer has a thickness of 5 to 30 nm. A variety of materials are used: those that reflect independent from the wavelength (mostly metals, TiO_2) and those that absorb certain wavelengths and reflect others (iron oxide, chromium oxide). In the second group, the interference color of the SiO_2 is superimposed by the absorption-color of the oxides.

Metal halogenides are not appropriate for the CVD-process for the coating of metal platelets with metal oxides, because they tend to react with the metal surface and furthermore, the steel reactor is corroded in the chloride atmosphere. In principle, $TiCl_4$ and steam can be used to manufacture TiO_2/Al-pigments [23], however, the process is very complex and to date has not been used in practice.

Mica pigments can also be coated with Fe_2O_3 using the CVD-process [23–25]. However, since effect pigments of the same composition produced using the process described in Chapter 2.8.2 are much less expensive, the fluidized bed reactor process has never been used in practice.

2.14 Multilayer pigments with a Fabry-Perot structure manufactured using the PVD-method

Since 1990, in addition to the optically variable pigments described in Chapter 2.12 and 2.13, products with a strong angle-dependent color effect that are produced using the PVD-method (Physical Vapor Deposition) have come onto the market. They are mainly based on a metal-dielectric, multi-layer structure and have not only a distinctive color travel, but also high color intensity. Different colors can be achieved by coordinating the film thicknesses and types of pigment used and carefully controlling the parameters. In order to ensure a minimum difference in color, the film thickness must be kept within a tolerance of a few atoms [18, 111].

The metal layers are often made up of chromium (a semi-transparent absorber) and aluminum (a reflector impermeable to light). Magnesium fluoride or silicon dioxide are used for the dielectric layers. Symmetric configurations as shown in Figure 2.54 are used for pigments, whereas large-area non-pigmented optical coatings can also be made from asymmetric layers. The chosen layer structures are the basis of an optical phenomenon known as the *Fabry-Perot* effect. Such multi-layer interference configurations can be constructed so that parts of the visible spectrum are strongly reflected, whereas there is little reflection in the neighboring wavelengths. Thus the requirements are met for very high color intensity. The special spectral properties of the pigments are maintained by the extremely exact adjustment of the film thicknesses.

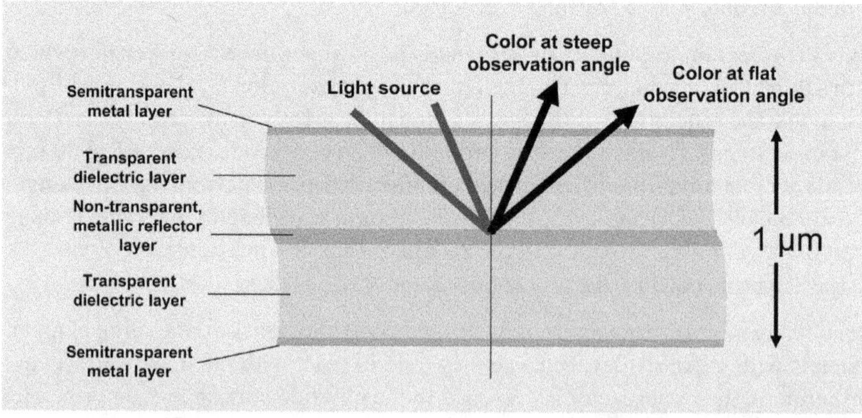

Figure 2.54: Schematic illustration of the layer structure of optically variable pigments (OVP) with Fabry-Perot structure

Platelet-shaped pigment particles that are particularly suitable for a strong angle-dependent effect need the symmetric layer configuration mentioned above. At least five layers are necessary. However, a significantly higher number of layers are sometimes required to achieve special coloristic effects. The necessary economic requirements for the production of multi-layer systems are difficult to achieve with the currently available vacuum-deposition technology. Therefore, only the five-layer system can be used for industrial purposes.

Fabry-Perot pigments are typically produced in a series of special coating installations (roll coaters) that are situated in vacuum chambers. A release layer is firstly deposited on a moving transfer foil. This layer is made from a soluble polymer or a salt and is soluble in solvents. Thus coatings that have been deposited on the foil at a later stage can be removed at the end of the process.

After the application of the release layer, the transfer foil is placed in the vacuum-coating equipment. Once the air has been removed, the first layer of semi-transparent absorber metal (e.g. chromium) is vacuum-deposited on the foil. Then the first dielectric layer (e.g. MgF_2) and the non-transparent metal layer (e.g. aluminum) are applied. Subsequently, the second dielectric layer and the second semi-transparent absorber layer are applied in the same way. Once the symmetric multi-layer configuration (e.g. Cr-MgF_2-Al-MgF_2-Cr) has been fully applied, the transfer foil is removed from the vacuum equipment. The multi-layer film is removed from the carrier by dissolving the release layer.

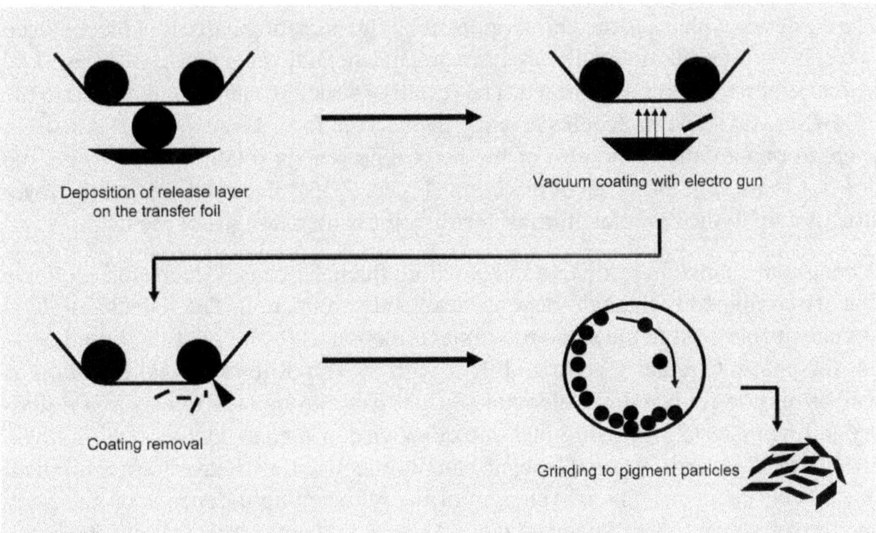

Figure 2.55: Process scheme for optically variable pigments with Fabry-Perot structure

In the following stage, the released coating components are ground and sorted to achieve the required particle-size distribution. The resulting multi-layer pigment particles have a diameter between 1 to 100 µm and are from 0.2 to 2 µm thick. The complete manufacturing process is depicted in a schematic form in Figure 2.55. Pigments with a *Fabry-Perot* structure are used in anti-counterfeiting, for example in bank notes, share certificates, visas and passports [111].

2.15 Effect pigments based on liquid-crystal polymers (cholesteric effect pigments)

Interference color effects that are angle dependent can be created using liquid crystal polymers (LCP) based either on layers or particles. Such effects can, for example, be created either with platelet-shaped particles, which themselves are made from LCP material or with platelets that are evenly coated with a cross-linked liquid crystal polymer in a chiral-nematic, or cholesteric, arrangement [18, 112–114].

Liquid crystals are organic compounds that are in a state between that of an isotropic liquid and an anisotropic, crystalline solid [11]. They consist of long, stick or disc-shaped molecules. Depending on the arrangement of the molecules, different types of liquid crystal can occur: nematic, smectic and cholesteric. In nematic phases the molecules arrange themselves in a long-range order. In the average, the molecules lie in the same direction. The molecules lie parallel to one another along the director. Nematic phases behave therefore like crystals with one axis [11].

The cholesteric phase is the chiral opposite of the nematic phase. It is mostly used in displays. If chiral molecules are arranged in a nematic host phase, staircase-like helical superstructures are formed. The resulting structure can be understood in the following way: The molecules arrange themselves from layer to layer at a slight angle to one another. One turn of the helix represents a rotation of 360°, i.e. the director is back in the original direction (Figure 2.56). The thickness of the layer structure up to the complete turn of the director is the pitch (p) of the helix.

A continual change in refractive index within the helix causes interference effects that are combined with angle-dependent selective reflection. The reflected light is circularly polarized in the same rotational direction as the helix of the liquid crystal line phase. Circularly polarized light with the opposite rotational direction as well as all non-reflected wavelengths of light pass through the transparent liquid-crystalline material. By using just one clockwise or anticlockwise rotating layer, 50 % of the light arriving on the object can be reflected. If both layers are combined, full reflection occurs. The wavelength of the reflected light depends on the pitch and the refractive index. The bandwidth $\Delta\lambda$ of the reflected light follows the equation $\Delta\lambda = p \cdot \Delta n$.

The intensity of the total reflection is strengthened by the reflection from the various layers of the helix. To obtain the maximum reflectivity of a polarized state at least six helices or a thickness of approx. 3 µm is required [112]. The most effective reflection is achieved at film thicknesses of up to10 µm.

Cholesteric materials are temperature sensitive and show thermochromic effects. This is because of the temperature dependence of the pitch of the helix and the refractive index [112, 115, 116]. The three-dimensional cross-linking of the layers of liquid crystal fixes the existing situation and temperature dependence vanishes. At the same time, the high-viscous state of the liquid crystals is transformed to a polymer network which permanently maintains its form.

Cholesteric liquid crystals can be used either as polymerized platelets or directly as fluid phase. If they are used as a fluid, after a very short time, the liquid crystals orientate themselves into their supramolecular structure when brought on a substrate. UV-cross-linking produces coatings which are high-lustrous and iridescent and have good adhesion. They are "pigment-free colors" i.e. in contrast to conventional materials, colored paints that contain no pigment. The binder and colorant are identical [18, 115].

Often, polymerized pigment platelets are made from polysiloxanes. A thin film of the liquid crystal is photochemically polymerized and subsequently broken down into platelets. The pigments are colorless and transparent. They demonstrate a

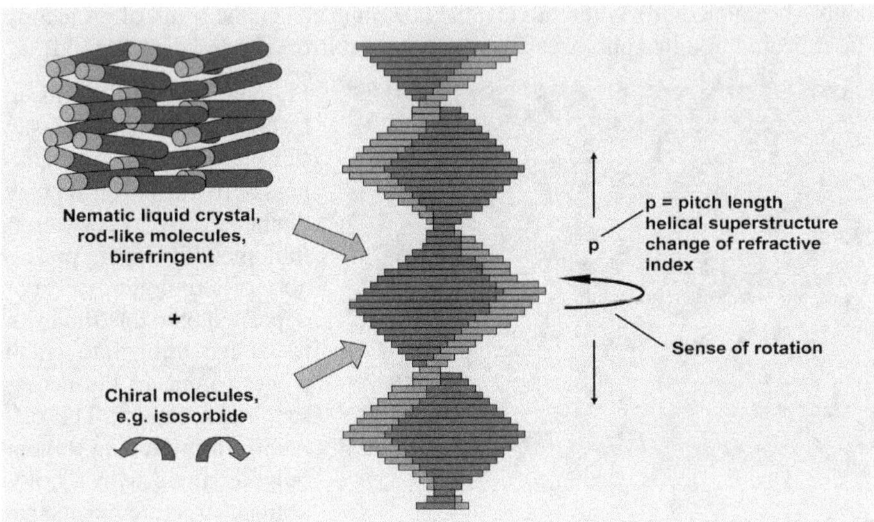

Figure 2.56 Formation of cholesteric liquid crystal using the reaction of a liquid-crystalline nematic phase with chiral molecules. The formed cholesteric liquid crystal shows a strong color travel due to multiple interference [114]

strong angle-dependent interference effect in a paint film especially when using a dark background. When the substrate is white, one sees merely a nacreous sheen, whereas on black areas, pure rainbow colors with a strong color travel become visible. Commercially available pigment particles are over 4 µm thick. To date, the use of cholesteric pigments has been limited. This is due to the thickness of the platelets and their insufficient stability and workability for certain uses.

2.16 Structured effect pigments

Whilst the color effects of all pigments described before have been due to the interference of light on thin layers, it is also possible to create special effects by the bending of light at surfaces with a regularly-repeating structure. Two types of structured pigments are commercially available: holographic and diffractive effect pigments.

2.16.1 Holographic pigments

Platelet-shaped pigment particles can be formed by crushing holographic films into small pieces [18, 117, 118]. Figure 2.57 shows the surface of a pigment particle with a holographic structure. Basically, any basis material that has a symmetrical, holographic structure can be employed to make holographic flakes. However, in practice, holographic mirrors are most frequently used. They are the simplest reflection holograms. They can be produced by splitting a laser beam and subsequently reuniting the divided beams at a photosensitive layer on a thermoplastic film (double-beam method). Alternatively, the hologram can be the result of projecting a laser beam through a photosensitive layer on a mirror (single-beam method).

Figure 2.57: Scanning electron micrograph of the surface of a holographic pigment. The average distance of the elevated positions in the structure amounts to 800 nm.

Pigment particles produced by crushing of thermoplastic films have a minimum thickness of 10 µm and an average diameter of 10 to 300 µm. A thin metallic layer, preferably of aluminum, is vapor deposited onto the film prior to its reduction into small pieces to ensure the necessary reflection [18]. The difficulty in producing thinner polymer films with a holographic structure means that their use has been restricted to the printing and paint sectors.

2.16.2 Diffractive pigments

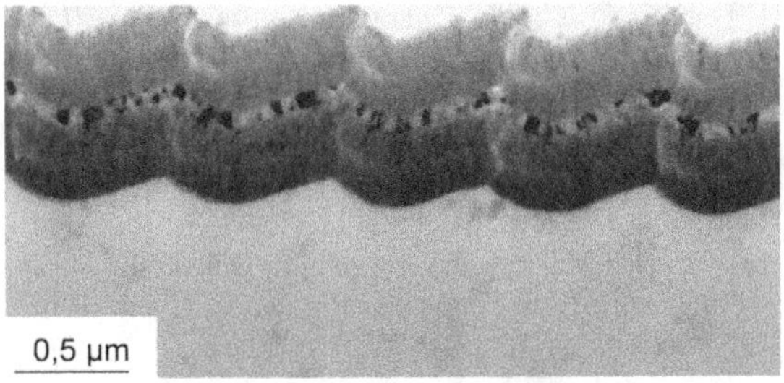

Figure 2.58: Transmission electron micrograph (cross-section) of a diffractive pigment. Layer structure: MgF_2-Al-MgF_2

Diffractive pigments are produced by vacuum deposition of thin films onto a structured-polymer film template. Typical examples of the order of layering on the polymer film are MgF_2-Al-MgF_2 or SiO_2-Al-SiO_2. There is a release layer between the film and the layers produced by the PVD-process (see Chapter 2.14). At the end of the deposition process, this release layer is removed. The resulting pieces of deposited material are then collected, broken down into particles and then fractionated [38, 119]. Their typical thickness is less than 1 µm. The periodic repetition of the structure is around 1 µm and it has a depth of several nanometers [119]. Figure 2.58 shows an image of such a diffractive pigment, taken with a transmission-electronic microscope. MgF_2 or SiO_2 act as mechanical support layers for the thin, structured aluminum film.

As the expensive structured film can be used only once and vacuum technology is needed, the production costs are extremely high. Thus these pigments are used only in such specialized areas as security printing.

References

[1] DIN 55943:2001-10, Farbmittel (Begriffe)

[2] Glausch, R., Kieser, M., Maisch, R., Pfaff, G., Weitzel, J., Special Effect Pigments, Vincentz Verlag, Hannover, 1998

[3] Franz, K.-D., Emmert, R., Nitta, K., „Interference Pigments“, Kontakte (Darmstadt) (2) (1992) 3

[4] Pfaff, G., „Perlglanzpigmente“, Chem. unserer Zeit, 31 (1997) 6

[5] Snow, M. R., Pring, A., Self, P., Losic, D., Shapter, J., „The Origin of the Color of Pearls in Iridescence from Nano-Composite Structures of the Nacre“, Amer. Miner., 89 (2004) 1353

[6] Handbook of Optical Constants of Solids, ed. E. D. Palik, Academic Press, Inc., Harcourt Brace Jovanovich Publishers, 1991

[7] Bäumer, W., “Perlglanz und Perlglanzpigmente”, Farbe + Lack, 79 (1973) 747

[8] Klug, A., Franklin, R. E., Biochim. et Biophys. Acta, 32 (1959) 203

[9] Sanders, J. V., „Colour of Precious Opal“, Nature, 204 (1964) 1151

[10] Luck, W., Die BASF, 13 (1963) 23

[11] Sage, I., Ullmann‘s Encyclopedia of Industrial Chemistry, 5th ed., Vol. A 15, VCH Verlagsges. Weinheim, 1990, 359

[12] Steinsträßer, R., Chemiker-Zeitung, 95 (1971) 661
[13] Hampel, B., Z. Werkstofftechn., 3 (1972) 149
[14] Marchmay, T. A., Sci. Amer., 185 (1921) 533
[15] Taylor, H. F., Drugs Oil Paints, 3 (1937) 106
[16] Hachisu, S., Sci. Light (Tokyo), 6 (1957) 20
[17] Greenstein, L. M., Proc. Sci. Sect. Toilet Good Assoc., 45 (1966) 20
[18] Pfaff, G., Reynders, P., „Angle-Dependent Optical Effects Deriving from Submicron Structures of Films and Pigments", Chem. Rev., 99 (1999) 1963
[19] Soder, G., DE 17 45 392
[20] Asahi Kasei Kogyo, DE 20 41 244
[21] Pfaff, G., Industrial Inorganic Pigments, 3rd ed., Wiley-VCH, 2005, 230
[22] Huber, A., Pfaff, G., "Spezielle Effektpigmente", Phänomen Farbe (3) (2005) 34
[23] Ostertag, W., "Effektpigmente", Nachr. Chem. Technol. Lab., 42 (1994) 849
[24] Ostertag, W., Bittler, K., Bock, G., EP 68.311
[25] Ostertag, W., XIXth FATIPEC Congr., Aachen, Bd. 2 (1988) 103
[26] Schröder, H., Kaufmann, R., US 3,138,475
[27] Pfaff, G., Andes, S., Bauer, G., Brenner, G., Brückner, H.-D., Heyland, A., Kuntz, M., Osterried, K., Schmelz, M., WO 97/43346
[28] Pfaff, G., Andes, S., Hock, S., Brenner, G., Brückner, H.-D., Heyland, A., Kuntz, M., Osterried, K., Schmelz, M., WO 97/43348
[29] Mattila, H. O., Eskelinen, P. J., EP 287 354
[30] Korpi, T. M., Hyttinen, S. J. O., Vapaaoksa, P. J., EP 325 484
[31] Vapaaoksa, P. J., Eskelinen, P. J., EP 377 326
[32] DeLuca, C., Kurtenbach, W. P., US 5,611,851
[33] Ohkura, K., Matsuzaki, S., Eskelinen, P. J., EP 666 292
[34] Bäbler, F., EP 604 370
[35] Bugnon, P., Herren, F., Medinger, B., EP 408 498
[36] Bäbler, F., Jaffe, E. E., EP 466 649
[37] Löbbert, G., Ullmann's Encyclopedia of Industrial Chemistry: Phthalocyanines, 6th ed. (electronic release), VCH Verlagsges. Weinheim, 1998
[38] Maile, F. J., Pfaff, G., Reynders, P., „Effect Pigments – past, present and future", Progr. Org. Coat., 54 (2005) 150
[39] Benbow, B. J., de Jong, B., Adams, J. W., Ullmann's Encyclopedia of Industrial Chemistry, 5th ed., Vol. A 16, VCH Verlagsges. Weinheim, 1991, 551
[40] Gysau, D., Füllstoffe, Vincentz Verlag Hannover, 2005
[41] Greenstein, L. M., Pigment Handbook, 2nd ed., Vol. 1, Wiley New York, 1988, 829
[42] Schmidt, C., Friz, M., „Optical Physics of Synthetic Interference Pigments", Kontakte (Darmstadt) (2) (1992) 15
[43] Pfaff, G., Perlglanzpigmente, Vincentz Verlag Hannover, 1995, 13
[44] Atwood, F. C., US 2,278,970
[45] Linton, H. R., US 3,087,828
[46] Linton, H. R., US 3,087,829
[47] Kohlschütter, H. W., Rössler, H., Getrost, H., Hörl, W., Reich, W., US 3,553,001
[48] Quinn, C. A., Rieger, C. J., Bolomey, R. A., US 3,437,515
[49] Kumar, K. N. P., Keizer, K., Burggraaf, A., Okubo, T., Nagamoto, H., „Textural evolution and phase transformation in titania membranes: Part 2 – Supported Membranes", J. Mater. Chem., 3 (1993) 1151

[50] Eskelinen, P., „X-Ray Diffraction Study of TiO_2 Thin Films on Mica“, J. Solid State Chem., 100 (1992) 356
[51] Eskelinen, P., Ritala, M., Leskelä, M., J. Solid State Chem., 103 (1993) 160
[52] Leskelä, M., Eskelinen, P., Ritala, M., „Thermoanalytical Studies on TiO_2-Mica Pigments", Thermochim. Acta, 214 (1993) 19
[53] Eskelinen, P., "Effect of SiO_2, NaCl, Al_2O_3, and $FeCl_3$ on Phase Change Behavior of Supported and Unsupported TiO_2", J. Solid State Chem., 106 (1993) 19.
[54] Hildenbrand, V. D., Doyle, S., Fuess, H., Pfaff, G., Reynders, P., "Crystallization of Thin Anatase Coatings on Mica", Thin Solid Films, 304 (1997) 204
[55] Pfaff, G., "Effektpigmente – Nano-Metalloxidschichten auf plättchenförmigen Partikeln", Praxis der Naturwissenschaften: Chemie, 55 (2006) 26
[56] Ostertag, W., Mronga, N., US 4,978,394
[57] Ostertag, W., Mronga, N., EP 338 428
[58] Thurn-Müller, A., Hollenberg, J., Cosmetics and Toiletries Manufacture, Century Press Bushey Heath, 1991, 72
[59] Thurn-Müller, A., Hollenberg, J., Liston, J., „Special Effect Pigments and Fillers for Cosmetics“, Kontakte (Darmstadt) (2) (1992) 35
[60] Emmert, R., Weigand, M., EP 307 747
[61] Maisch, R., Pinture e Vernici, 66 (1990) (4) 9
[62] Schmidt, C., Maurer, U., "Interference for the New Millennium – Timiron Splendid Colors", Euro Cosmetics, 8 (1999) 28
[63] Schmidt, C., Bernhardt, K., Schoen, S., Heinz, D., US 6,599,355
[64] Pfaff, G., "Anorganische Pigmente – Aktueller Stand und neue Entwicklungen", Phänomen Farbe (7/8) (2002) 19
[65] Franz, K.-D., Ambrosius, K., Wilhelm, S., Nitta, K., WO 93/19131
[66] Hauf, C., Kniep, R., Pfaff, G., „Preparation of Various Titanium Suboxide Powders by Reduction of TiO2 with Silicon“, J. Mater. Sci., 34 (1999) 1287
[67] Ostertag, W., Mronga, N., Hauser, P., "Eisenoxidbeschichtete Aluminiumpigmente", Farbe + Lack, 93 (1987) 973
[68] Franz, K.-D., Ambrosius, K., Prengel, C., EP 354 374
[69] Franz, K.-D., Ambrosius, K., Knapp, A., Brückner, H.-D., US 4,867,793
[70] Pfaff, G., Reynders, P., "Neue Entwicklungen bei Metalloxid-Glimmer-Pigmenten", Chem. Rundschau Jahrbuch, 1993, 31
[71] Pfaff, G., Glausch, R., Maus, W., XXIIth FATIPEC Congr., Budapest, Bd. 1, 1994, 56
[72] Kimura, A., Suzuki, F., Nkamo, M., Fukushima, S., J. Soc. Cosmet. Chem. Japan, 21 (1987) 238
[73] Esselborn, R., Bernhard, H., Hesse, R., Russmann, H., US 3,926,659
[74] Emmert, R., „Instrumentation Color Matching of Pearlescent Pigments“, Cosmet. Toiletries, 104 (1989) 57
[75] Watanabe, T., Noguchi, T., US 4,435,220
[76] Franz, K.-D., Rau, A., US 4,545,821
[77] Franz, K.-D., Ambrosius, K., Knapp, A., Brückner, H.-D., EP 246 523
[78] Noguchi, T., EP 339 399
[79] Watanabe, T., Noguchi, T., US 4,603,047
[80] Pfaff, G., "Funktionelle anorganische Pigmente", Phänomen Farbe (7/8) (2003) 21
[81] Brückner, H.-D., Glausch, R., Maisch, R., "Neuartige helle, leitfähige Pigmente auf Glimmer/Metalloxid-Basis", Farbe + Lack, 96 (1990) 411

[82] Vogt, R., Neugebauer, E., Pfaff, G., Stahlecker, O., „Bright Conductive Pigments with Layer Substrate Structure", Eur. Coat. J. (7/8) (1997) 706

[83] Glausch, R., Pfaff, G., Maisch, R., XXIst FATIPEC Congr., Amsterdam, Bd. 2, 1992, 33

[84] Hocken, J., Griebler, W.-D., Winkler, J., "Elektrisch leitfähige Pigmente zur antistatischen Ausrüstung von polymeren Beschichtungsstoffen", Farbe + Lack, 98 (1992) 19

[85] Daponte, T. L. F., Verschaeren, P. C. M., Kieser, M., Edler, G., WO 94/05727

[86] Pfaff, G., Rosenberger, S., RAPRA-Conf. „Special Effects in Plastics", Berlin, 2002, Proceed., 41

[87] Pfaff, G., "Leserlich durch Laserlicht", Chemie heute (1998/99) 60

[88] Klenke, E. F., Stratton, A. J., US 3,087,827

[89] Franz, K.-D., Ambrosius, K., Brückner, H.-D., DE 41 25 134

[90] „Minatec": eingetragener Warenname der Firma Merck KGaA, Darmstadt

[91] „Lazerflair": eingetragener Warenname der Firma Merck KGaA, Darmstadt

[92] „Solarflair": eingetragener Warenname der Firma Merck KGaA, Darmstadt

[93] Merck KGaA, Effect Pigments for Plastics, Darmstadt, 2003

[94] Pfaff, G., High Performance Pigments, ed. H. M. Smith, Wiley-VCH, 2002, 77

[95] Nitta, K., Shau, T. M., Sugahara, J., EP 763 573

[96] Teaney, S., Pfaff, G., Nitta, K., "New Effect Pigments Using Innovative Substrates", Eur. Coat. J. (4) (1999) 90

[97] Pfaff, G., "Innovationen bei Effektpigmenten", Welt der Farben (1) (2000) 16

[98] Sharrock, S., Schül, N., „New Effect Pigments Based on SiO_2 and Al_2O_3 Flakes", Eur. Coat. J., (1/2) (2000) 20

[99] „Xirallic": eingetragener Warenname der Firma Merck KGaA, Darmstadt.

[100] Pfaff, G., Huber, A., "Spezielle Effektpigmente auf Basis von Aluminiumoxid-Plättchen", Welt der Farben (9) (2005) 14.

[101] Rüger, R., Oldenburg, N., Schulz, E., Thurn-Schneller, A., „Star Effects from Borosilicates", Cosmet. Toiletries (5) (2004) 133

[102] „Miraval": eingetragener Warenname der Firma Merck KGaA, Darmstadt

[103] Bauer, G., Osterried, K., Schmidt, C., Vogt, R., Knieß, H., Uhlig, M., Schül, N., Brenner, G., WO 93/08237

[104] Pfaff, G., „Special Effect Pigments Based on Silica Flakes", Inorg. Mater., 39 (2003) 123

[105] Schmid, R., Mronga, N., Radtke, V., Seeger, O., „Optisch variable Glanzpigmente", Farbe + Lack, 104 (1998) 44

[106] Wheeler, I., Metallic Pigments in Polymers, Rapra Technology Ltd., 1999

[107] Besold, R., Industrial Inorganic Pigments, 3rd ed., WILEY-VCH, 2005, 252

[108] Wißling, P., Metalleffekt-Pigmente, Vincentz Verlag Hannover, 2005

[109] Ostertag, W., Mronga, N., „Properties and Application of Luster Pigments", Makromol. Chem. Macromol. Symp., 100 (1995) 163

[110] Etzrodt, G., Kittel - Lehrbuch der Lacke und Beschichtungen (Pigmente, Füllstoffe und Farbmetrik), 2. Aufl., S. Hirzel Verlag Stuttgart-Leipzig, 2003, 160

[111] Phillips, R. W., Bleikolm, A. F., „Optical Coatings for Document Security", Appl. Opt., 35 (1996) 5529

[112] Makow, D., „Color Gamut of Liquid Crystal Polysiloxanes", Mol. Cryst. Liq. Cryst., 123 (1985) 347

[113] Robbie, K., Brett, M. J., Lakhtakia, A., „Chiral sculptured thin films", Nature, 384 (1996) 616

[114] Haremza, S., "Wenn Farben floppen - Effekte mit Flüssigkristallen", Welt der Farben, (3) (2000) 14

[115] Makow, D., „Reflection and Transmission of Polymer Liquid-Crystal Coatings and Their Application to Decorative Arts and Stained Glass", Color Res. Appl., 11 (1986) 205

[116] Tamaoki, N., Parfenov, A. V., Masaki, A., Matsuda, H., Mater. Res. Soc. Symp. Proc., 488 (1998) 915

[117] Boyle, E., "Fast & Flawless", Paper, Film & Foil Converter (May 1) (2004) 1

[118] Fitch, J. J., Fotlano, R. A., Josephy, K., Meikka, F. N., Meikka, R. G., EP 678 074

[119] Argoitia, A., Witzmann, M., Society of Vacuum Coaters, 45th Annual Techn. Conf. Proc., 2002, 539

3 Special effect pigments in paints

Dr. Frank J. Maile

3.1 General comments on the use, application and processing

This section is concerned with the use of special effect pigments in paints and coatings. The various technologies considered here are an application focus of this pigment class.

As mentioned in Chapter 2, materials referred to as "special effect pigments" are principally platelet-shaped luster pigments, with the exception of bronze powders and metal effect pigments. The pigments with their layer structure are based in most cases on natural or synthetic carriers, so-called substrates. Therefore the terminology "substrate-based effect pigments" is often used [1].

Before dealing with the subject in detail, it is worth considering why effect pigments are used in coatings [2, 3]. Many branches of industry, including paint manufacturers, focus their attention on the end users of their products to ensure that development work meets the needs of the market [4]. Furthermore, through the direct customer influence on color and design, the "time to market" should be kept as short as possible. This situation exists across the wide area of industrial design, fashion, architecture and the automobile and branded-products industries. New product development is virtually always prompted by the end-users' constant search for individuality or uniqueness in new products [5].

This demand can be met by the use of effect coatings to create clear-cut product differentiation and is extremely important for the future of the effect pigment industry [6–8].

Advances in research and development as well as in the industrial production of coloring agents in the 19th century brought synthetically produced pigments and dyes onto the market and thus enabled color to be put into everyday products [9]. An important tool for creating differentiation had arrived.

With the introduction of metal effect pigments in the 1920s, a new family of coloring agents appeared, which has since become very popular in the automotive industry [2, 10]. These platelet-shaped, metal effect pigments give a light to dark flop effect depending on the angle from which they are observed. The anisotropy of the platelet-shaped particles and the high reflectivity of the metal surface have led them to be frequently described as miniature mirrors. The morphology and size distribution of the particles, allow surface effects such as a coarse appearance and pronounced texture to be achieved. A comprehensive review of current developments in this field can be found in [2].

A further milestone in the development of coloring products was the introduction of mica-based special effect pigments in the 1970s. They can be used to give a clear-cut differentiation between painted industrial products. Although the platelet shape is similar to the metal effect materials mentioned above, the two groups differ in their chemical and physical properties. A detailed treatment of the various types of substrate-based effect pigments can be found in Chapter 2.

Most substrate-based effect pigments have a transparent support layer that has the anisotropic morphology of a platelet [11]. During the pigments' development in the second half of the 20th century, it became clear that muscovite mica was the most appropriate material for use as the substrate (Chapter 2). Today, mica is still the most commonly used substrate for transparent and semi-transparent effect pigments. The crystal structure of the alumosilicate muscovite mica layer is crucial for the performance of effect pigments based on it (see Chapter 3.4).

It should also be mentioned that the term mica is very often used in the context of automotive paints to mean the whole family of mica-based effect pigments. From a scientific point of view, this is not completely accurate, because, as described above, the term refers only to the support and not to the complete effect pigment. It would be more correct to use the term metal oxide/mica effect pigment [12].

Another expression often used in the paint world is pearl. This is an abbreviation for pearlescent and denotes an optical effect: nacreous luster. This technical term describes the optical (semi-) transparency of substrate-based effect pigments in their final state and thereby underlines its important distinction from non-transparent, metal effect pigments [13]. The materials which are best suited for use on a substrate such as mica are principally transparent or at least semi-transparent and highly refractive metal oxides such as TiO_2 (anatase, rutile) and α-Fe_2O_3 (hematite) [14]. In general, the color of this family of pigments is formed by the overlapping of light waves that have been reflected by the layers of pigment. The resulting color can be controlled by varying the type of pigment layer, its chemical composition and its thickness (Chapter 2). To enable interference to occur, the light waves must be coherent. Interference in coherent light waves occurs when light is reflected by parallel layers as in pearlescent pigments; at the same time the pigment geometry and chemistry described above must obey certain physical requirements in order that interference may occur [15, 16].

The following principles for the use of these coloring materials in paint formulations can be concluded [1]:

- Substrate-based effect pigments demonstrate the phenomenon of interference that occurs with transparent or semi-transparent materials. With decreasing transparency, there is a reduction in both interference properties and effect (see Chapter 3.8).

- Substrate-based effect pigments always have a lower hiding power than metal effect pigments.
- The required transparency must be considered during formulation: the prescribed covering power at a particular film thickness must be adhered to. This interrelation needs to be respected when substrate-based effect pigments are combined with transparent organic or inorganic pigments, carbon black and/or metal effect pigments, or used in so-called tri-coat paint systems, i.e. colored primer/basecoat with effect pigments/clear coat (see Chapter 3.8) [17].

Articles or information on styling, color matching and formulation using coatings which contain effect pigment are rare [18]. It should be mentioned here that formulation expertise is often company property that is the result of many years of endeavor and is protected by rigorous secrecy agreements. It is in many ways part of a paint manufacturer's capital.

This is because of its use in automotive paints, repair paints, coil coatings, powder coatings etc. These are areas that require different technologies (solvent-based/ water-based/high solids) or application methods (high-rotation atomization/ spraymate, tribo/corona etc.). The successful use of these technologies makes specialist knowledge indispensable. This expertise has very often been built up from very long series of practical, "trial-and-error" tests over many years.

This empirically obtained knowledge, built up with much effort over a long period of time is carefully protected. Often this knowledge is not published (even in the form of patents). On the other hand, guidance formulations now exist which can be used as starting points. Given their source (often manufacturers of additives, binders, or pigments, who seek to use these formulations to promote their products), they become less sensitive information [19–21].

The following is the definition of an effect coating:
According to the German industrial specification DIN 55945 (1999), an effect coating is one where, in addition to color, visually perceptible properties such as luster (as opposed to a glossy surface), observation angle color dependency, structure or texture are present.

The paint industry today can call on a wide range of inorganic and organic pigments, with a broad spectrum of colors and effects. Thus suitable products are available to meet a wide range of demands. In practice, as stated above, the strengths of the individual product families are frequently combined by using mixtures of organic and inorganic pigments. This often offers a cost advantage [22].

When choosing an appropriate coloring agent, it must not be forgotten that the profile of a product's properties does not depend solely on the coloring agent. The form, distribution and alignment of the particles and their interaction with the polymer matrix as well as with additives determine color, effect, hiding power and weathering properties [22]. Thus, special effect pigments are important, but are only one of the building blocks which determine the quality of an effect coating. For this reason the closer one looks at the various classes of raw materials, the more it becomes apparent that they are made up of a large number of individual substances [23]. As an example, the range of available binders involves besides natural materials, which are still widely used, a large and steadily growing number of synthetic products. Here there is a clear trend towards water-based coatings products that can be applied under conventional application conditions [22].

It can be seen in Chapter 2 that an increasing number of effect pigments are being developed using carefully defined synthetic substrates. A large and steadily increasing variety of inorganic and organic coloring agents are on the market, which have a known chemical structure and can be reliably produced in large quantities. Development work is focused on the synthesis of new effect products which open up new areas of color. At the same time, they strictly comply with all the technical and health-and-safety standards governing their handling in the paint production plant, during their subsequent use and at a later stage in the recycling process [22].

Attention is being given to additives, plasticizers and solvents with regard to the physiological aspects of their use as they affect the workplace and environment [24]. Whilst in the 1920s paint solvents were seen as indicators of technical and economic advance, by at the latest 1974, the adoption of the German Federal Emission Act had noticeably changed attitudes. The maxim of current and future activities is either the reduction of levels of solvent or their replacement by physiologically harmless water or solvents that have been shown to be harmless. Solvents which can be built into a film such as the reactive thinner for radiation-hardened paints are a further focus for future paint development [22].

To sum up, it can be said that, taking into consideration all work and environment-related factors, high quality is shaping the development of all the raw materials used in paint, including special effect pigments [22]. This has resulted in paint producers setting clear requirements for the makers of effect pigments. The criteria thus formulated must be fulfilled in demanding situations in particular in the so-called high-performance area [2, 25]:

- Regulatory measures
- Safety (safe handling of pigment powder)
- Coloristic effects (interesting color location/effect, reproducibility)
- Cleanliness (particle size distribution/over-coarse grain)
- Technology (freedom from agglomeration, re-dispersibility, humidity resistance and stone-chip resistance, adhesion etc.)

- Processability (degree of dispersibility in a medium, settling properties, durability, rheological stability circulation-system test)
- Application properties (hiding power, cloudiness, haziness, anti-sagging properties
- Weathering stability (Florida, accelerated weathering)
- Measurability (in the paint film)
- Repairability (on the production line as well as in repair body shops)
- Color harmony (match to mounted parts)
- Innovation
- Acceptable price
- Global availability/just-in-time availability etc.

3.2 Technology comparison for the coloring of plastics and paints

Mass-colored plastics in form of masterbatches (see Chapter 4) and film technologies are being considered at present in various publications as alternatives to painting processes [26, 27]. Thus, at this point a short comparison of the two areas will serve as a guide and highlight the important differences between the two processes.

Mass-colored plastics are manufactured by the incorporation of pigments into melted polymers during their conversion into a granulate. This is referred to as the masterbatch. This granulate is a premix for use in the pigmentation or coloring of finished products (for example by injection molding) in the plastics-processing industry [10].

The advantages of coatings over mass coloring are the reflective properties (gloss and brilliance) and the desired surface homogeneity and uniformity that the painting process achieves. However, surface defects can arise in the application of paint. These can sometimes be rectified only by expensive repair processes.

The appearance of paint films which contain platelet-shaped, substrate-based effect pigments depends on the shape of an orientation distribution curve which depicts the variation of the orientation of platelets in the dried film relative to the substrate (0° = parallel to the substrate). Furthermore, if the distribution of the effect pigment particles over the painted surface is irregular and visible differences in the concentration of particles occur, the paint surface appears to be "cloudy".

As previously mentioned, the gloss and brilliance of a paint film are two major advantages over mass-colored plastics. This can be seen when the differing platelet-orientation distribution curves for painted surfaces and mass coloring are compared (see Chapter 3.6).

Virtually regardless of the method of application, coatings demonstrate a greater level of homogeneity in visual effect, assuming of course that the material is correctly applied. Many visible surface defects such as flow lines or welding seams are hidden. In this respect, the best result is achieved with spray application. Dipping is less satisfactory as is brush application (see Chapter 3.6). This is because an uneven and inhomogeneous application caused by "unpracticed" brush marks lead to a cloudy appearance. In the case of dark, muted colors these problems are less visible, because the absorption pigments present in the recipe supply the necessary hiding power.

Color flop is a characteristic phenomenon of all platelet-shaped effect pigments. It is the varying perception of color as the viewing angle changes [13, 28]. Detailed information on this can be found in Chapter 2. It should be noted that color flop is a characteristic of coatings that cannot be accurately reproduced in the mass-coloring of plastics and is a further advantage of coatings [10].

The application temperature window already mentioned restricts the range of uses of a large number of pigments. This is why many of them can be used only in paints. For example, many organic pigments show limited thermal stability. Thermally stable pigments for use in mass coloring are much more expensive and thus represent a cost disadvantage.

Temperature stability does not in general present a problem for substrate-based effect pigments, because apart from surface-treated WR-pigments (WR = weather resistant, see Chapter 3.5), they are made of metal-mica compositions that have been subjected to temperatures above 600 °C and thus display the thermal properties of these products. This also applies to special effect pigments on other substrates used in practice.

The advantages of mass coloring lie in the coloristic properties, processing, environmental compatibility and cost. The stability of colored, mass-pigmented articles is an advantage over colored, relatively thin paint films where, for example delamination can be a problem. This is particularly relevant for parts of a car body that are near the ground, such as wheel caps, body trim etc. Furthermore, mass coloring can be advantageous for complex shapes, where spray application, due to the form of the spray pattern and the geometry of the spray mist can lead to an uneven distribution of paint. On the other hand, during the forming process, an object can experience sudden differences in viscosity. This can result in flow lines, which can disfigure the surface. Here a more uniform appearance can be achieved by painting.

With regard to the process steps of the individual techniques, prior to the painting process a substrate to be coated must undergo meticulous preparation, in particular when dealing with polymeric products [29, 30]. These materials often have poor wetting properties. A good example is the polyolefin family. They have low surface energy, which makes wetting difficult. Thus pre-treatment processes such as flame treatment or the use of surface-activating chemicals need to be considered to ensure adequate adhesion of the coating. This has resulted in the development of a remarkable technology to deal with the problem. It is based on silane, titanate and zirconate adhesion promoters (used by, for example, Kenrich Petrochemicals, Inc.) [31].

Often paints containing substrate-based effect pigments are applied in two coats: a 10 to 20 µm layer of basecoat containing the effect pigments, followed by a 25 to 40 µm clear coat layer to provide protection and gloss (see Chapter 3.8). In contrast, mass coloration is carried out in one step during the processing of the polymer.

Environmental restraints on the use of solvents, the resulting emissions and the disposal of paint waste are putting companies that use coating materials under increasing pressure [32]. For example, the EU-Solvent Directive requires that solvents may represent only 30 % of the annual EU-emission of 12 million tonnes of VOCs (VOC = Volatile Organic Compound). Since 1990, VOCs have been cut by 70 to 80 % as part of the drive to reduce the level of ozone in the atmosphere to an acceptable level. The disposal of paint waste in landfills has meanwhile become subject to an extra tax based on the weight of material involved. Future legislation will probably be even stricter. Although in recent years there have been enormous strides in the introduction of water-based technology, the investment costs for installing corrosion-resistant equipment are still very high. In the mass coloring process almost no waste or overspray occurs. What is more, incorrectly molded parts or extruder waste can be collected and recycled either as the same or other products.

The cost aspect is very important. Indeed it is often the deciding factor for success in a competitive situation. Both the investment and running costs of a painting plant and drying equipment are considerable. Also the repair of paint-film defects by sanding and repainting is labor-intensive and therefore expensive. This is why the interest in mass-coloring technology has been rising recently. This tendency would have been greater if it were not for the immense problems associated with achieving a result which is comparable to that of a painted body part as mentioned above. Furthermore, the savings of mass coloring are often cancelled out by the higher raw material costs occasioned by the organic pigments needed to formulate chromatic shades, which must be temperature stable.

3.3 Incorporation, dispersion and stabilization

The processing of pigments, for example by dispersion, is a separate area within the field of paint research and technology [33–35]. Before dealing in this section with the dispersion of effect pigments, the important factors influencing it should be mentioned. *Kaluza* divides the treatment process in four stages [33]:

1. The dispersion stage, i.e. the incorporation of the pigment powder in the medium to be colored.

2. The stage that represents the liquid material, after dispersion, but before application.
3. The stage when the liquid material is applied on a substrate for example by spraying, dipping, rolling etc.
4. The stage when the pigmented paint has been cured and is dry.

Regardless of their coloristic properties, the most important characteristics of pigments in paints can be summarized as follows:

5. pigment concentration
6. packing density of the primary particles and the aggregates in the agglomerations
7. pigment surface polarity
8. chemical nature of the pigments
9. particle size and distribution (PSD)
10. shape of the particle (morphology)

The main properties of the medium that effect the pigment processing are [36]:

11. binder concentration
12. binder viscosity
13. solvent type
14. chemical nature of the polymer
15. real molecular size, calculated using the molecular weight, degree of cross-linking, solvation and association
16. shape of the molecule (morphology)

The interrelationship between the four main stages and the parameters that influence them are depicted by *Kaluza* [33] as a flow diagram and illustrate the complexity of the situation.

The particle geometry, surface properties and processing of special effect pigments make them significantly different to inorganic white or colored pigments. In general, they are easier to work with. Figure 3.1 uses REM images to underline the differences in the particle morphology and size of industrially produced titanium dioxide pigments. As a comparison, the spots visible in the image on the left correspond to the average diameter of the type of titanium dioxide shown on the right. As rule of thumb it can be said that typical effect pigment particles are 100 to 1000 times larger than those of conventional titanium dioxide. Thus sedimentation occurs more frequently in liquid paint systems in which effect pigments are used (see Chapter 3.4) [37].

When a pigment is used, information on two properties is always needed: the oil absorption value and/or the BET specific surface area. Oil absorption is affected by various properties of a pigment powder, for example, the particle size and specific surface area. The smaller the particles or the larger the specific surface area,

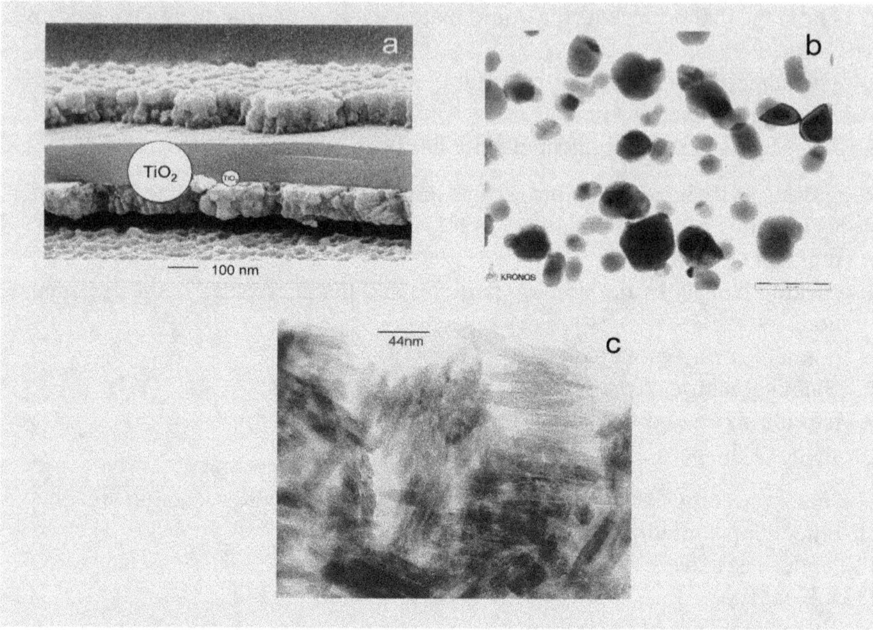

Figure 3.1: Scanning electron micrographs of various types of titanium dioxide to demonstrate the relative differences in particle size: (a) cross section through a typical, industrially produced titanium dioxide mica effect pigment (rutile), (b) industrially produced titanium dioxide (rutile), (c) industrially produced nano-titanium dioxide (rutile). As a comparison, the spots in (a) correspond to the average diameter of the type of titanium dioxide shown in (b) and (c).

the more oil or binder will be needed to create a continuous, adsorptive coat around the particle [38]. Furthermore, the particle packing density of the powder plays a role. The lower the packing density or the greater the pore volume, the more oil will be needed, not only to create the coat round the particle, but also to fill the pores and interstices. Finally, the flow properties of the pigments in the medium under test are important. These depend not only on factors such as PSD, on morphology, packing density and pigment concentration in the binder, but also binder viscosity. The energy factors between the constituents of the mediums and the particle surface play a role here; factors which in turn have an effect on the flow properties [33].

It can be said that the oil absorption values of effect pigments are "moderate" when compared to many inorganic pigments. Depending on the type of substrate, the particle size distribution, or the post processing and weather stabilization of the pigment, typical values lie between 30 to 90 g/100 g.

However, the situation is different for the BET-surface area of substrate-based effect pigments [39]. BET is the specific total external and internal surface of dispersed or porous solid matter, expressed as the amount of gas adsorbed by a body using the method developed by *Brunauer*, *Emmet* and *Teller* (BET method) [40].

As previously described in Chapter 2, various substrates such as mica, synthetic silicium dioxide, synthetic aluminum oxide and borosilicate glass are used for the manufacture of a substrate-based effect pigment. The special feature of the method is the precipitation of defined and reproducible thin layers of metal oxide that are neither irregular nor show side precipitation. The metal-oxide layer produced in the process is nanocrystalline and microporous. The observed specific surface area of these nanocrystalline layers lies between 30 to 120 m^2/g despite a relatively large particle diameter [41, 42].

The dried products are not normally sufficiently chemically inert for use in paints and thus need to be thermally treated to give them the required light stability and an extremely low level of reactivity. During this treatment, as in ceramic systems, compaction and granulation occur in the nano-crystalline layer. Taking titanium dioxide layers as an example, after calcination above 600 °C, the columns of nano-crystals turn into granules, which can be as large as the film is thick (see Figure 3.1, left). A lateral displacement of material does not take place. The granules are separated by mesopores and micropores are no longer present. The specific surface area thus attained is around 10 m^2/g.

The surface properties of substrate-based effect pigments are such that dispersion in liquid paints is regarded as "easy". This means that, given the correct expertise, they can be used in almost every paint system [43].

The following method is used for the incorporation of the pigments into a liquid paint system, whereby, due to differences of a historical nature, the process can vary in practice from one paint maker to another.

- The effect pigment is dispersed and wetted under agitation
  Depending on the paint system involved, alcohols, ethers, ether esters or aromatic hydrocarbons are used at a pigment to solvent ratio of 1:1 to 1:2. Sometimes special wetting agents are added to dissolve any agglomerates present. This process is helped by the size and shape of the particles.
- Addition of the binder under agitation
  The agglomerates that are still present can be broken down with a dissolver, that is often equipped with a special disc (e.g. the Lenart from Vollrath, see Figure 3.2, page 102).This takes place at a low rotation speed (approx. 1 m/sec. which allows for the "doughnut effect" without foaming) and for a short operational time (10 to 15 minutes).

The dissolver should be used "as little as possible, but as long as necessary". Under no circumstances should the rotation speed exceed the above figure otherwise the effect pigment will be damaged due to the friability caused by the thinness of the particles. Methods using a higher shear strength (such as ball mills, free-roll mills and bead mills) will very probably damage particles and thus lead to a subsequent change in the particle size distribution (Figure 3.3). The resulting increase in the number of edges will increase the light diffusion and thus cause a reduction in luster.

- Stabilization of the dispersed state [44]
  Here, appearing problems have less to do with possible flocculation, but more with settling and the subsequent reincorporation of the sediment as well as, in the case of water-based paints, the pH and viscosity [45].

An increase in viscosity of the liquid paint slows down sedimentation. However, the viscosity specified for the process must be respected. Electrostatic stabilization using carboxyl additives or those containing an amine group works by causing repulsion of similarly charged particles. If the pigment surface is treated with long chain organic molecules, the probability of the occurrence of settlement that is difficult to reincorporate is reduced. Which of these methods is better to use (alone or in combination with others) depends on both the effect pigment and the binder and can be determined only empirically [43].

Figure 3.2: Photograph of a Lenart disc (diameter 65 mm); it is used for the incorporation of effect pigments in a liquid paint using a dissolver or mixer

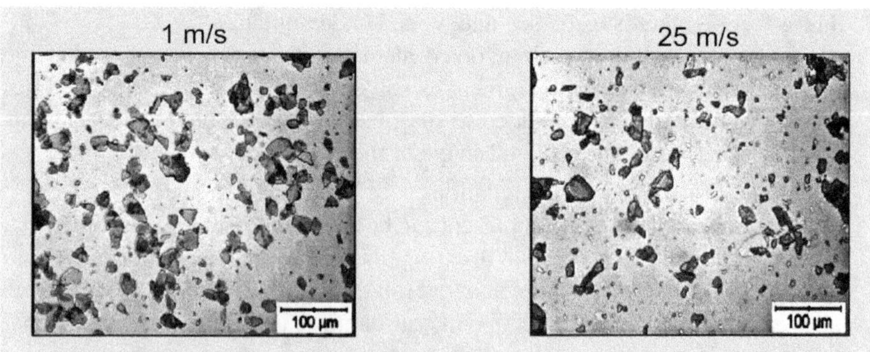

Figure 3.3: Bright field microscopic analysis of paints containing titanium dioxide mica effect pigments after differing degrees of shear stress caused by a laboratory dissolver equipped with a dissolver disc: rotation speed 1 m/s (left), 25 m/s (right); a comparison of the images confirms the damage caused by a rotation speed of 25 m/s.

The special, stabilized metal effect pigments that are needed to prevent the graying and gassing problems that occur when metallic luster pigments are used in water-based systems, are not needed for special effect pigments.

3.4 Sedimentation

The importance of sedimentation in paints containing special effect pigments is described in the following [46].

The composition and structure of synthetically produced, substrate-based effect pigments has been thoroughly described in Chapter 2 as well as in various review articles [13, 17]. The important aspects are summarized below:

- pigment particles are anisotropic platelets
- typical mean diameter of the particles is between 5 and 100 µm
- particle thickness is less then 1 µm

In comparison, inorganic white pigments, for example, are almost spherical in form with a diameter of <1 µm and thus considerably smaller (see Figure 3.1). The reason for this is clear: In order to achieve maximum coverage, the particle size must be that which gives maximum diffusion of light [47]. In special effect pigments, however, light diffusion must be kept to a minimum so that the typical change in perceived color with changing observation angle can take place. The pigment layers are built up of substances with a high refractive index (titanium dioxide, hematite). They also have a high specific gravity and thus increase the weight of the resulting pigment. As a result, suspensions of effect pigments differ from normal, colloidal dispersions in that the large particles in the former induce anisotropic, interparticular attraction.

The actual settling process takes place in two steps:

- The particle settles according to Stoke's law. However, the viscosity, polarity and specific gravity of the solvent exert an important influence on the process [48, 49].
- The resulting sediment compacts under the strong influence of interparticular attraction [50].

It is not possible to differentiate strictly between the two processes. In particular in very concentrated suspensions, large agglomerates made up of many particles can be formed during the settling process [49].

Settling

As special effect pigments are larger and heavier than normal, colloidal particles, gravity outweighs Brownian motion and drags the particles to the base of the medium, where they form a more or less compact sediment [49]. Thus, suspensions

*Table 3.1: Rate of settling calculated according to Stokes [46], pigment density d_s 3.1 g/cm^3 ("Iriodin" 100), viscosity η = 0.952 mPa*s (water at room temperature), height of 10 cm and K from [51, 52].*

Average particle diameter [µm]	Rate of settling calculated according to Stokes
0.1	264 years
1	2.6 years
10	9.6 days
100	2.3 hours

of these pigments tend to settle over a period of several days. The rate of settling can be estimated using Stokes` law assuming the particles to be spherical and at a low concentration:

$$v = K \cdot \frac{r^2 (d_s - d_l)}{\eta}$$

Here K = constant, d_s = specific gravity of the particle, r = particle radius, d_l = specific gravity of the medium, η = medium viscosity [51, 52].

According to Stoke's law, pigment settling is influenced by the following factors:

- properties of the pigments (size and PSD, specific gravity, morphology and concentration)
- properties of the medium (specific gravity, viscosity, polarity)
- additives (the surface interaction of the pigment and medium, wetting, etc.)
- thermic or mechanical vibrations (Brownian movement, temperature, storage conditions)

Taking "Iriodin" 100 of Merck (titanium dioxide/mica) as an example of a special effect pigment, Table 3.1 shows the calculated rates of settling in water. Due to the lack of data for other particle shapes, the spherical form was used. Results showed that, rates of settling ranging from weeks to hours can be expected for a PSD of 5 to 200 µm.

Compacting

The compacting of mica-based, special effect pigments is assessed by looking at the anisotropic distribution of electrostatic charge. The faces are negatively charged whereas the edges carry a positive charge. There are two borderline cases: Situation a) where the inter-particular reaction is fully developed and situation b) where it is suppressed (Figure 3.4).

Situation a)

Here a loose, soft and voluminous sediment with a structure similar to a "house of cards" is formed by the electrostatic attraction of the positively charged edges and negatively charged faces [53–57]. The sediment can be broken up by a weak force [58]. This situation is comparable to an unstable or flocculated colloidal dispersion. The liquid above the sediment is cloudy [51, 52, 59–63].

Situation b)
In this case, the sediment formed is dense, hard and compact. This occurs when the effect of the electrostatic charge on the pigment is cancelled out by protonation or adsorption and the sediment is compacted by gravity. This results in stacks of platelets [64]. As there is no attraction between the particles, this situation corresponds to a stable, colloidal dispersion of small particles [49]. During the settling process no aggregates are formed and the liquid above the sediment is clear [49, 59–62, 65].

One would normally assume that the strong force of attraction in situation a) would cause the density of the sediment to rise. However, this has not been observed either in the mica-based effect pigments that have been tested [46] or in ceramic materials which have no anisotropic charge distribution [53–57, 59–62].

In point of fact, if the attractive force is strong enough, the particles that collide quickly stick together. However, the probability is low that the particles that come into contact with the agglomerate present will adhere at the position that leads to the densest packing. Thus the result is an agglomerate with the open structure that is desirable for good reincorporation.

When working with effect pigments, it is quickly clear whether one is dealing with situation a) or b). Situation a) is preferable as it leads to voluminous, loose sediment, which can be easily re-dispersed in the medium, thus ensuring homogeneous consistency. In contrast, in the case of situation b), the user has to solve the problem of breaking up a hard cake of settled paint. This leads to the further problem that

Figure 3.4: Illustration of the compacting process of mica-based effect pigments. There are two different borderline cases: full development (left) and suppression (right) of the interparticular forces [46].

re-dispersion by stirring or mixing is almost impossible. It poses the interesting question: how can a sediment be forced to have the consistency of situation a)?

As long the resulting sediment of an effect pigment can be easily reincorporated, it can be regarded as a side effect of the use of this special class of pigment. The re-dispersibility of these pigments varies according to their properties (surface treatment or weathering stability, PSD) as well as those of the medium (pH, salt concentration and solvent polarity) [46].

The sedimentation of metal oxide-mica pigments occurs in two phases: the settling process (duration: minutes to hours) and the concurrent compacting process (duration: weeks to months). The mass of the individual particles is such that the settling of effect pigments is unavoidable. The only way to reduce this is by increasing the viscosity, i.e. by fixing the pigment particles in a solid matrix. One method is to use bentonite or hectorite to form a gel with a yield value [66]. Another possibility is the use of a rheology additive which operates on the polymer thickener principle [67, 68]. The degree of success of each measure depends, however, on the paint technology involved. Thus every case must be treated separately.

The consistency of the sediment of mica-based effect pigments is due to the phenomenon of electrostatic interaction. This either helps or hinders the formation of a house-of-cards structure. The following conditions promote the formation of the desired soft sediment:

- water-based medium with a low pH, or better
- non-polar solvent (see Figure 3.5)
- moderate salt concentration

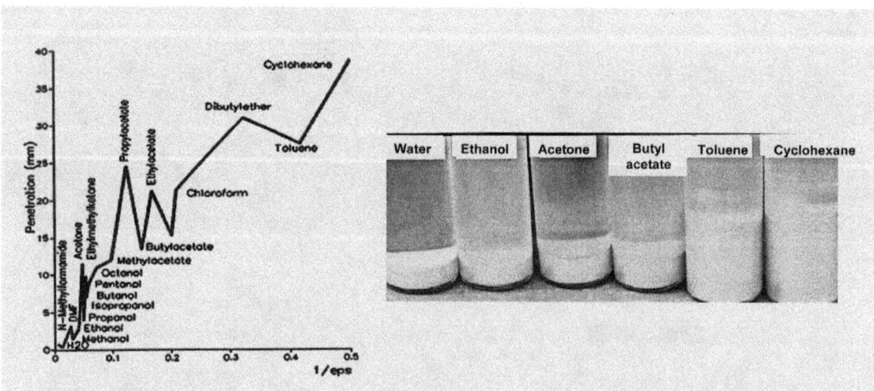

Figure 3.5: Illustration of the penetration depths of "Iriodin" 100 (titanium dioxide mica effect pigment) sediments measured at 10 pond (Dynometer) in different solvents as a function of the inverse dielectric constants (left) and the corresponding images (right) after four weeks of sedimentation [50]

- large difference in electrostatic charge between the flat surfaces and edges of the mica particles
- high isoelectric point

Re-dispersibility is more difficult in suspensions in water-based mediums than in organic equivalents owing to the nature of commercially available, water-based technologies [69]. Finally, it should be borne in mind that additives and surface-active substances can have a considerable effect on the sediment. This is the subject of current research [70].

3.5 Surface treated effect pigments in paints

An important task of surface treatments and weather stabilization is to suppress the catalytic activity of the metal oxide used. The following explanation is focused on titanium dioxide (particularly in the anatase form) which displays special properties (it is a semiconductor) [47, 71]. Weather stabilization needs to take place on the pigment surface, as this is where the catalytic process happens. The surface should be completely covered with catalytically inactive material. Protective layers are typically less than 10 nm thick. Their chemical composition enables them to act as a redox buffer on the pigment surface. Organic materials are also used as a protective layer to control the wetting of the pigment by the medium [72].

Another important task of surface treatments or weather stabilization is to ensure compatibility between the strictly inorganic metal-oxide surface and the organic medium (polymers). In particular, the polarity and wetting are modified and the tolerance can be improved [23].

Surface treatments and weather stabilization are, on one hand state-of-the-art technologies, on the other they are part of the important expertise of paint makers and are, like information such as paint formulas, they are kept secret. The pigment manufacturers offer surface treatments that are made to measure for their customers' systems and technologies. This often involves them in joint projects [25]. In view of the complexity of paint technology and the individual characteristics of customers' systems, it does not make sense for a pigment manufacturer to carry out developments alone. On the other hand, such specially developed materials lead to a complex range of products. If you add to this the paint manufacturers' demands listed in Chapter 3.1, it is clear that the ability constantly to deliver material in the required quality represents a significant challenge for the pigment supplier.

Surface treatments have an important influence on the performance of special effect pigments in paint systems [74]. The reasons for the surface treatment of pigments are: the particle size and their size distribution, the intrinsic surface chemistry and the tension in the polymeric paint film which after curing can be seen as an inorganic-organic composite.

Here polymer domains are formed in the macromolecular paint film, which function as a homogenous layer and show cohesive/adhesive properties. The presence of pigment particles, in particular large ones, can destroy this homogeneity and lead to a weakening of the film structure.

Surface chemistry can also play a role, because the particles have polar surface functionality. Water is omnipresent, mobile and polar and can cause interaction through hydrogen bonding. Furthermore, it can diffuse through polymer films and deposit itself on the effect pigment/paint film interface. Subsequently, diffused water can become fixed by hydrogen bonding and, as it accumulates, act as a wedge which leads to separation of the pigment and film. This can result in a sometimes irreversible increase in the L*-value and can give blisters and delamination.

Another factor is tension in the film. This can be due to residual tension from the film-building (coalescence) and drying processes. In addition, tension can be caused by the swelling of the basecoat due to the absorption of water.

This residual internal stress in the film is caused by, among other things, its thermic history. As a cured film cools, the paint film and substrate shrink to a different degree due to the different expansion coefficients. Further tension occurs at the interfaces, because the effect pigments cannot deform sufficiently. In extreme cases, this can cause micropores, which in turn allow humidity to enter. If the uptake of water causes swelling, the bending stiffness increases significantly [75, 76].

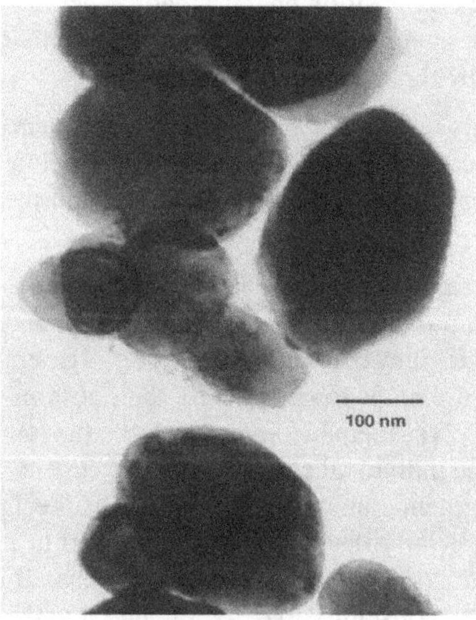

Figure 3.6: Transmission electron micrograph of DuPont "TiPure" R706, TiO_2 cores (dark crystallites) with surface treatment. The denser components of the surface treatment are visible, but with a relatively low level of contrast; the amorphous part of the surface treatment is hardly visible [109].

Looking at the swelling of an un-pigmented paint film, relatively little tension is induced, as the forces pull in the opposite direction and can act freely. When large, platelet-shaped pigment particles are present, shear forces are also induced on the platelet surfaces. This causes them to rotate and thus cause disorientation.

Effect pigment particles can also be the cause of other types of forces as they are "water barriers" which prevent water from moving under the platelets. This situation causes differential swelling and thus leads to unequal forces within the film. This can then give distortion of the paint film and a subsequent loss in distinction of image (DOI). These factors can lead to a loss of adhesion, particularly if the film is subjected to moisture.

As mentioned above, the performance of effect pigments is greatly affected by surface treatment. It is assumed that hydroxyl groups interact more strongly than the oxide groups with the polymer. This happens for example when there is greater interaction as well as a higher number of bonding sites.

Silanes are well known as adhesion promoters for many paint systems. This also applies to effect pigments [77]. The individual functional groups of silanes react with every available site, both with oxides and hydroxides on the surface as well as among themselves. In the latter case, polymer groups are formed. Thus, organic residues can enter into various acid-base or hydrolysis reactions. Naturally these are stochastic processes, the frequency of which depends on the reaction conditions (concentration, temperature, pH etc.).

The "surviving" functional organic groups are able to interact with the film former. At the same time, the organic network around the effect pigment reduces its water affinity, because the presence of the organic material reduces the surface energy.

Figure 3.6 shows a TEM image (TEM = transmission electron microscope) of a TiO_2 pigment which has undergone surface treatment.

3.6 Particle geometry and orientation

The unique visual effects of special effect pigments can be understood as the consequence of chemical composition and the physical and optical properties associated with it (see Chapter 2). A comparison of their particle size and geometry as well as their specific gravity with that of other inorganic and organic coloring agents emphasizes the differences already mentioned. Effect pigment particles are bigger by a factor of a hundred to a thousand. This can lead to problems in their use. Another difficulty associated with the size and density of effect pigments is the settling described in detail in Chapter 3.4.

The anisotropic morphology of the particles influences optical parameters such as gloss, interference color, texture or sparkle. Changes in the paint technology or process parameters have an effect on the platelet orientation [78, 79].

The major advantage of – or difference with – so-called electrostatic application (ESTA), or "high-rotation atomization" as it is referred to in the literature, is the more or less complete separation of the atomizing process and the transport of the droplets. This is the reason why it is easier to control the spray-jet geometry using the so-called deflection air and electrostatic field forces than is the case with other application methods [80].

Microscopic studies of cross sections show the orientation of the effect pigment particles in the paint film. They confirm that the application process has a decisive effect on particle orientation within the polymer matrix. The changes in the way orientation takes place, that can be analyzed with optical and electron microscopes, make themselves visually apparent in the optical parameters mentioned above such as color and gloss.

The reasons for these variations are as follows: (1) Separation of the particles during and after application and before the completion of film coalescence (darker = electrostatic application); (2) a higher proportion of "disoriented" particles in an electrostatically-applied layer in comparison to one applied pneumatically. This leads to significant changes in remission properties (see "differences in lightness") that should be measurable. To verify this, remission curves of the individual surfaces have been measured with an ETA multi-FX10 instrument. The results showed the expected differences.

Application processes of liquid paints used in industry (see Chapter 3.7) need, in the case of pneumatic application, to take the following parameters into account, because they also affect the particle orientation in the paint layer:

- automatic spray equipment
- spray guns
- nozzle diameter
- air pressure (nozzle, etc.)
- spray gun – object distance
- cross coat
- number of spray cycles
- flash-off time between the spray cycles
- application viscosity etc.

These parameters are responsible for:

- geometry of the paint droplets
- size distribution of the paint droplets
- distribution curve of the spectrum of droplets
- droplet collision speed
- shape of the spray cloud that hits the substrate

Figure 3.7 clearly illustrates the factors that influence the film formation of a water-based paint layer [81].

The presence of platelet-shaped effect pigments in polymer droplets hinders their coalescence and thus the paint-film formation. This is another factor which influences the standard deviation of the particle orientation in the polymer matrix. The evaporation/emission of solvents [82], or in the case of water-based paints of co-solvents, as well as stratification [75, 76, 83] of the polymer film are also important factors.

All these factors taken together explain why, for example, the repair of automotive paints should be undertaken by experienced sprayers. The process is a real challenge as all the application steps are critical [84].

One way of influencing the orientation of lamellar particles in a paint film is by using special additives in the form of solid materials in the paint recipe. An example is spherical polymer particles (see Figure 3.8, left at the bottom) [79, 85].

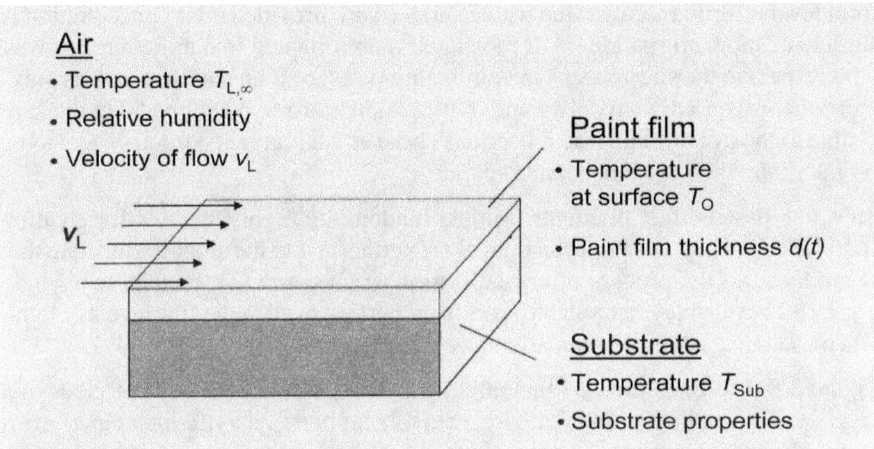

Figure 3.7: Illustration of a water-based paint film during drying and the factors that affect film formation [81]

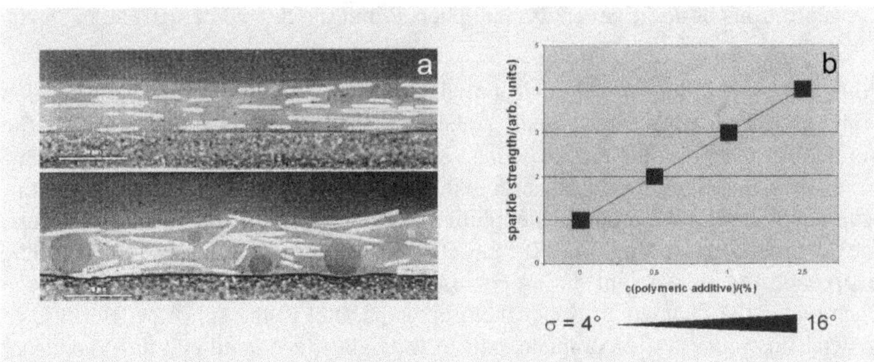

Figure 3.8: Orientation characteristics of effect pigments in layers of car-repair basecoats, (a) solvent-based, acrylic-melamine materials without (above) and with (below) spherical polymer particles, (b) "sparkle strength" (arbitrary units calculated from the visual assessment of panels) versus the concentration of polymer spheres as well as standard deviation σ of the pigment particle orientation (0° = platelet is oriented parallel to the surface of the painted substrate, ± x° = tilt of the pigment particle (clockwise/anti-clockwise direction) [110].

This additive is often used by repair-paint manufacturers to correct properties such as color flop and sparkle, but in point of fact, its origin lies in the need to create a perfect match on repaired parts or areas of a car on the automobile production line.

The term "sparkle" describes the ability of platelets to reflect light in various directions and is closely associated with the size and orientation of the effect pigment particle [86, 87]. Platelets which all lie relatively parallel to one another in a paint film, for example, the "leafing" type, reflect light in one direction [10]. They show a uniform level of brilliance over the whole surface and, provided their concentration is sufficient, show no sparkle [87]. If the platelets are oriented in a more random way with respect to the substrate, as in non-leafing types, or if an additive such as polymer spheres is used, then only some of the platelets are so positioned that they are visible to the eye through true reflection. These effect pigment particles seem to be brighter than their neighbors and "sparkle".

Substrate-based effect pigments using corundum are highly suitable for creating sparkle [86, 87]. They show special optical properties due to the smooth substrate that is produced in the process of crystallization (each corundum platelet is a single crystal). They display remarkable sparkle properties, even under low levels of light, when used in a suitable formulation (see Chapter 3.8).

Figure 3.8 illustrates the way that effect pigment particles orient themselves in a layer of automobile repair paint basecoat (solvent borne, acrylic-melamine), with and without the presence of polymer additives (spheres). Their influence on the orientation can be clearly seen in the cross-section (Figure 3.8 bottom left). A correlation can be deduced between the disorientation, here caused by the various concentrations of additives (polymer spheres) and the degree of sparkle, or "sparkle strength" [86, 87].

Another way of influencing the orientation of effect pigment is to use materials which contain either leafing or non-leafing platelets. Leafing effect pigments cover the surface of an appropriate matrix with layers of overlapping particles. The light scattering from the edges is reduced because the lamellar particles are well oriented. Non-leafing effect pigments contain oleophilic additives. In contrast to leafing pigments, they exhibit good wetting. Thus the platelets are evenly distributed in the paint. The evaporation of the solvent during coalescence and the subsequent film shrinkage encourages the platelets to orient themselves parallel to the substrate. Because in general this process is incomplete, part of the light is scattered and thus the visual impression given is one of "sparkle" rather than an even, reflecting surface.

On the other hand, solvent-free powder coatings demonstrate reduced orientation in comparison to solvent or water-based paint systems (Figure 3.9). This is due to the absence of the process described above, which helps the platelets orient themselves parallel to the substrate. Thus uniform reflection is very limited due to the fact that the particles are statistically randomly oriented.

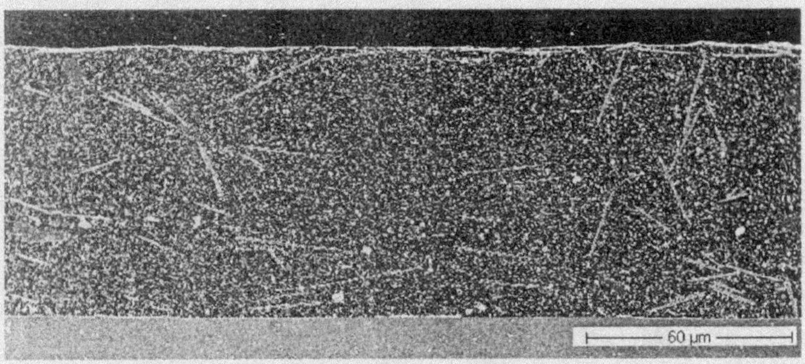

Figure 3.9: Scanning electron micrograph (FIB preparation of the cross section), which shows the orientation of effect pigment particles (5 % of an industrially prepared titanium dioxide mica effect pigment, silver-white type, RAL 7037 formula) in an approx. 80 µm layer of a powder coating ("Primid") [110].

The concept of leafing and non-leafing cannot be applied to the mass-coloring technology described in Chapter 3.2, because the melting viscosity of the polymers is too high to allow leafing to occur. Further differences between "painting" and "mass-coloring" technologies are discussed in Chapter 3.2.

3.7 Consequences of the application

Effect paints, i.e. those containing platelet-shaped pigments, can be applied using various methods, for example spraying, rolling, dipping or brushing.

Spray application
Effect paints can be sprayed without problem. However, the various spray methods lead to differences in the effect that is developed during application. This must be taken into account for each method.

- Air-assisted spray
  When the air pressure is sufficient (4 to 5 bars), the effect development is good. Under the correct conditions, a good spray pattern can be achieved. This method is used for effect paints in both the industrial and automotive paint sectors.

- Airless spray
  Effect paints can also be applied using the airless technology. A high circulating-system pressure is needed to replace the missing air pressure in order to ensure an optimum pigment orientation. As in general the wet film thickness is higher

than for air-assisted spray, the effect development is not as good. This method is mainly used in the industrial-paint sector and has limited use for effect paints.

- High-rotation atomization or electrostatic application (ESTA) [79]
  High-rotation atomization is increasingly being used for ecological reasons. The big advantage is that the so-called "overspray", i.e. the quantity of sprayed paint that does not land on the object being painted, can be greatly reduced [80]. The result is less organic solvent in the exhaust air. Paints containing effect pigments can be applied well using this method. The effect development mechanism has been described in [88].

High-rotation atomization is used particularly in the automotive industry for metallic and pearlescent paint application. The first coat is applied electrostatically and the second using air-assisted spray. The sequence high-rotation atomization/pneumatic, which is also called ESTA/Spraymate or bell/Spraymate in the automotive industry, makes repair work easier, both on the production line and in repair body shops [32].

Rolling

Rollers are generally used to apply dispersion paints containing effect pigments for use on mineral or organic substrates. Sufficient pressure must be guaranteed in the whole system throughout the application process in order to achieve the optimum particle orientation needed to ensure a good effect and appearance. Depending on the nature of the rollers, undesirable striping can sometimes occur.

Dipping

In the dip-painting process, external force is not needed to achieve good pigment-particle orientation. This is helped by the following measures:

- The use of a low-solids recipe. The high level of shrinkage causes the orientation.
- Application of the paint in several coats. Every layer should be as thin as possible (approx. 5 μm) so that the platelet-shaped effect pigments are automatically oriented by the flow process.

In general dipping is used in the industrial-paint sector, as well as for specialized situations such as the coating of costume jewelry such as pearls and, the painting of three-dimensional parts.

Brushing

Application by brushing is a problem area for effect paints, because the pigment orientation changes with each brush stroke. This leads to a surface with a structured appearance. A better effect can be created only by choosing a more appropriate paint system. The situation is similar to dip-paints; good effect development can be achieved only when a binder is used that has high shrinkage during the drying process.

"Old" application methods

Manual techniques such as stamping, filling/puttying or "dot spraying" with glazing compositions can be used with effect pigment containing dispersion paints to create sophisticated, aesthetic as well as imaginative, glittering and brilliant effects [89–92].

3.8 Color and effect creation

General comments

Ranges of mica-based effect pigments contain luster pigments with colors and effects ranging from silverwhite, interference colors and copper to red, gold and green. By combining these with other transparent coloring agents, the user and color designer have a wide spectrum of color creation at their disposal.

In general, the effect pigment concentration ranges from 1 to 4 wt-% of effect pigments and inorganic or organic coloring pigments are used at the same level. These figures are for materials at application viscosity.

The coloring pigment is first pre-dispersed to its maximum color strength. Then the effect pigments are added. The use of coloring pigment compensates for the low hiding power of, in particular, silverwhite and interference pigments. In certain cases, various effect pigments may be used together to create new color effects. When various interference pigments are combined, it must be remembered that the color mixture is additive. In the case of complementary colors this leads to a grey-white color.

Effect creation

At this point, it is worth giving some examples of the large number of different effects that can be achieved today by using diverse types of effect pigments:

- metal oxide/mica effect pigments: pearl luster effect (e.g. Merck's "Iriodin" pigments)
- metal oxide/mica effect pigments: sparkle (e.g. Merck's "Xirallic" pigments)
- metal oxide/silicium dioxide effect pigments: multi-color effect (e. g. Merck's "Colorstream" pigments)

The formation of the individual effects in paints is influenced by a number of factors:

- type of effect pigment
- the combination of effect pigment/organic/inorganic coloring agents
- pigment fraction used (size of the pigment particles)
- pigment concentration
- nature of the coating layer
- nature of the substrate (shape, surface)
- application method

The properties inherent in the above factors are looked at below:

Types of effect pigment
The combination of effect pigment/organic or inorganic pigment

During the formulation of a color, the main limiting factors to be taken into consideration are the transparency of the effect pigment and the required hiding power. This is, for example, very important when a mica or metallic color is being prepared for the automotive industry. This leads to a number of general principles.

For base coat/clear coat application, it is impossible to achieve sufficient coverage with a product that contains only effect pigments. If nevertheless, a pure effect pigment appearance is desired, then a high-opacity product, usually a white solid color, must be first applied on the filler. This is known as a tri-coat system.

The following solutions are possible to formulate effect pigment containing products with sufficient hiding for base coat/clear coat application:

- Use in combination with carbon black. This naturally leads to darker, deeper colors.
- Use in combination with transparent organic or inorganic pigments. Here deep, brilliant colors can be formulated. Depending on the color, carbon black may need to be added.
- Sometimes it helps to use effect pigments with a smaller particle size, i.e. with a greater ability to scatter light, to achieve the necessary coverage. However, under no circumstances may an arbitrary increase in effect pigment level be used to improve the opacity.
- Another solution is to add a small amount of aluminum pigment. The desired hiding is quickly achieved, but, as is the case with effect pigments with smaller particle sizes, it is at the expense of brilliance, effect and chroma.

In practice, all four solutions are used. But the choice depends on which method allows the formulation of colors with sufficient hiding to be achieved. Thus, a typical automotive "mica-coating" is made up of the following components: effect pigment, transparent organic or inorganic pigment, carbon black and possibly aluminum pigment. The ratio of the individual components can vary. Approximate values are: approx. 60 % effect pigment, approx. 35 % organic or inorganic pigment, approx. 2 % carbon black and a maximum of 3 % aluminum.

Pigment fraction used (particle size distribution)
Metal oxide-mica pigments with a particle size of approx. 10 to 60 μm (normal fraction) show the typical pearl luster effect.

Fractions with smaller particle sizes, i.e. 5 to 20 μm, or with particle sizes in general under 15 μm, increase the hiding power of the coating by light scattering, but their gloss and brilliance are less. Good gloss and brilliance are achieved by using significantly larger particles, but at the cost of hiding. A similar situation occurs when special effect pigments are used on other substrates.

It is possible to achieve brilliance and good coverage, within certain limits, by mixing effect pigments of the normal and small particle fractions.

Pigment concentration

The optimum pigment concentration for the formation of a pearl luster effect lies between 1 and 4 wt-% at application viscosity. The proportion of effect pigment in relation to the binder is normally 10 to 25 wt-%. Up to 40 wt-% may be used to achieve special effects or if the paint film thickness is very low.

Higher pigment levels are inadvisable, as they hinder the orientation of the platelets. This in turn leads to a reduction in gloss and effect formation. Sometimes, clear coats can be livened up by adding 1wt-% (based on the binder) of effect pigment.

Pigment concentrations needed for multi-color effects and sparkle

Similar concentrations of pigments may be used to achieve multi-color effects (for example with metal-oxide/silicium dioxide effect pigments, "Colorstream") as are employed when metal oxide-mica effect pigments are used to produce a pearl luster effect. However, sparkle needs very low effect pigment levels (typically 0.5 to 1.5 % at application viscosity) (see Chapter 3.6).

Composition of the coating layer

The composition of an effect coating layer is as follows:

The substrate is painted first with a primer and then with a filler. This is followed by a pigmented base coat and then normally a layer of clear coat.

The effect formation is influenced by

- color and roughness of the filler
- hiding power of the base coat
- thickness of both the base coat and clear coat

Figure 3.10 (page 118) shows a typical structure of a base coat/clear coat effect application as well as the layer sequence in automotive paints.

Nature of the substrate

The nature of the form to be painted has a pronounced influence on the effect formation, in particular the reflection properties. For example, a rounded form shows the reflection and color flop properties better than a flat one.

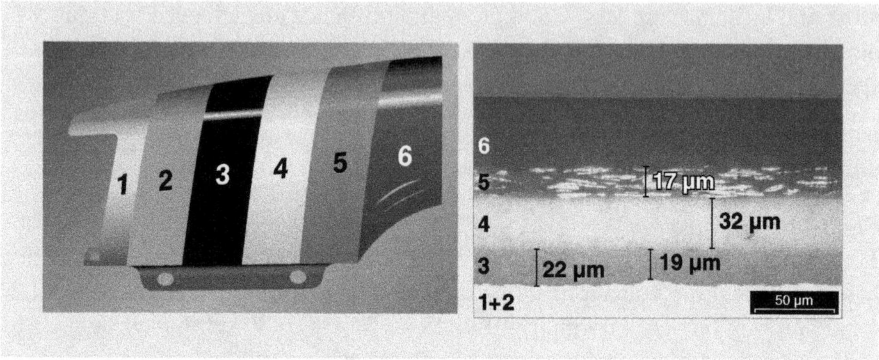

Figure 3.10: Structure of an automotive basecoat/clear coat effect coating with the corresponding layer sequence [22]

Details of the application parameters

Application parameters such as spray pressure, nozzle size, distance between the nozzle and object to be painted, number of coats applied as well as, in the case of electrostatic, high-rotation atomization, the voltage and bell rotation speed, have a decisive effect on the creation of the effect. Thus much attention must be paid to these parameters [80].

3.9 Special effect pigments in paints

Effect pigments can be used with all commonly available binders and polymer resins such as acrylics, polyurethanes, epoxy resins etc. The platelet orientation and therefore the luster effect depend on the solids level of the paint.

General comments on the behavior of special effect pigments in paints

The high level of solvent in a low solids system (weight solids <30 %) promotes good orientation, because of the large amount of shrinkage that takes place. Polar solvents encourage the breaking down of agglomerates and the wetting process. This is also the case for medium solids paints (weight solids from 30 to 60 %).

Good platelet orientation is more difficult in high solids paint (weight solids above 60 %) [93]. However it is possible to improve the situation by increasing the wetting of the pigments. This can be done by the use of appropriate additives [94].

Orientation is most difficult in powder coatings and radiation-cured paints. However, the use of appropriate additives or specially prepared pigments can help achieve the required degree of orientation.

The advantages of the additives are as follows:

- They cover and thereby seal the complete pigment particle and thus adapt the properties of the pigment surface to the polymer paint system.
- They form a "pseudo-bonding" between the polymer in the binder and the pigment surface, i.e. the pigment and binder become alike at their interface. At the same time, the encroachment of water onto the pigment surface is prevented, which if it takes place has a negative effect on the visual effect formation.

In the high performance areas such as for automotive paints or facades, it is essential that effect pigments, are given surface treatment and weather stabilization (see Chapter 3.5).

Special effect pigments in water-based paints
One of the major tasks of the development of environmentally-compatible paint systems is the reduction of the emission of organic solvents [95]. In the following the special features of these paint systems with respect to the use of effect pigments are described.

In general, effect pigments can be easily incorporated into water-based paints. A small amount of pre-wetting with polar solvent, for example butyl glycol or water, helps the distribution of the pigments. However, in many instances, dry pigment can be incorporated successfully.

Water-based paints differ from their solvent-based counterparts, with regard to pigment incorporation and application, in the following ways [69]:

- Once the pigment has been added, pH and viscosity must be adjusted to the value for the paint in question. The pH is adjusted with a substituted amine such as diethanolamine or ammonia. Polyurethane or bentonite is used mostly as thickener.
- During application, the specified exterior temperature and humidity must be adhered to. These figures are also dependent on the material being applied.

Water-based effect paints in general have good stability. For some paint systems, however, the pigment surface must be modified with an appropriate surface treatment in order to optimize the performance of the pigmented paint.

Special effect pigments in powder coatings
The development of the modern, almost emission-free powder coatings began in the 1950s [96]. Since then, the demands made on these materials have grown continuously. Although the manufacturing and application processes have been optimized over the years [97–99]. They are not best suited for the use of effect pigments, including metal effect pigments. However, in recent years, there have been significant advances in improving the quality of formulations of powder coatings containing effect pigments.

The platelet shape and particle size of effect pigments require a special technique for use in powder coatings. In order to understand this, it is necessary to be acquainted with their normal manufacturing process and application possibilities.

Manufacture and application

The first step in the production process is to mix the raw materials, in their correct ratio, in a pre-mixer. This mixture is then melted in a double screw extruder. The material is thereby intimately mixed and the desired dispersion of the pigments is achieved. The extruded melt is cooled to room temperature and then broken up. The resulting small pieces of solid coating are then ground under conditions which give particles between 10 and 80 μm.

Reactive powder coatings are applied electrostatically. The material receives its electrostatic charge in the spray gun either by passing through a corona zone or by use of the tribo-electric effect.

Once having left the spray gun, the paint particles follow the lines of the electric field and are attracted to the earthed object to be painted. The charged powder particles induce an opposite charge on the object. This enables them to adhere to its surface.

Electrostatic application of powder coatings on a non-conductive material such as ceramic is not possible without help, because the adhesion by electrostatic charge is insufficient. The object must be first heated so that the powder coating immediately melts and thus sticks to the surface.

Quality aspects of a powder coating/pigment mixture

The particle size of effect pigments and powder coatings are approximately the same. Thus, if effect pigments are introduced directly into the premixing process prior to extrusion, the particles undergo extensive mechanical damage and lose most of their visual effect properties.

Effect pigments can therefore be used only in the so-called dry blend process. But even then problems are caused in conjunction with electrostatic application. Powder coatings and effect pigments do not separate in a predictable way. This means that during the usual process of reusing powder coating overspray, variations in color or luster can occur.

In practice, it is quite likely that the level of a particular effect pigment can vary before and after application. For example, a material which initially contains 5 % of effect pigment can have a level varying from 3 to 10 % after application onto a metal panel.

This separation problem is a longstanding concern of manufacturers of powder coatings and luster pigments [100]. It can be solved by, for example, using a thermic-mechanical process method in which the pigment particles are more or less completely fixed onto the powder coating particles [100, 101].

The process can be reproduced well in the laboratory. However, it is difficult to scale it up reliably to production levels. The temperature of the complete mixing room must be very even and under no circumstances may it exceed the maximum temperature specified for the powder coating. Mechanical stress on the effect pigment must also be kept within narrow limits to avoid batch to batch variations in effect. This treatment of powder coatings that contain luster pigments has become known as "bonding" or the "bonding process".

Special technical process aspects

All the methods used to solve problems which occur in the application of conventional solid powder coatings can also be used with materials containing effect pigments. The following "solutions" are of particular interest:

- When the highest possible quality of finish is desired, overspray should not be reused ("paint loss accepted").
- The use of overspray for other purposes.
- The normal solution is the continuous automatic addition of overspray to the original powder.

Film formation and the resulting effect formation take place during the stoving subsequent to application. The luster effect achieved in this way does not completely satisfy the expectations that one has from a solvent-based, low solids system. This is due to the less than ideal particle orientation.

An appropriate modification of the pigment surface (for example with saturated fatty acids) can significantly improve the orientation of effect pigments and thereby the luster effect [102]. These pigments show a light leafing effect, which allows the appropriate appearance to be achieved, even with highly pigmented powder coatings.

Alternatively, luster effects and color flop can be improved by modifying the powder coating [103].

Powder coatings are largely used for exterior paints. Thus it is important to choose effect pigments that have sufficient weathering stability. These are the so-called "weather-stable pigments" described in Chapter 3.5. Non-stabilized pigment variants are only recommended for use in interior paints.

The level of effect pigments used in powder coatings can vary considerably. In some cases 1 % is sufficient. The normal amount is 3 to 5 %, but 10 % and above may be needed to achieve a specific desired effect.

Special effect pigments in radiation-cured systems

Many classes of pigment absorb radiation and thus have limited use in radiation-cured systems. In contrast, effect pigments show sufficient transparency in a wide electromagnetic frequency range and can be widely used in this technology.

In this process only thin coats of material should be applied. The film thickness should not exceed 200 µm. It is unimportant whether the metal oxide used in the substrate-based effect pigment is TiO_2 and/or Fe_2O_3 or Cr_2O_3 [103, 104].

3.10 Special effect pigments in paint films and their properties

As with all pigmented paints, effect pigments must undergo tests prior to use to ensure that the pigmentation does not have a negative effect on the paint quality and, indeed, if possible, should improve it. The testing criteria are similar to those for normal pigments. There are, however, small differences. Both a high resistance to external influences and the stability of the luster effect are scrutinized.

Durability of effect-pigmented paints

Humidity stability

The humidity stability of effect-pigmented paints is measured using the so-called DIN 50017 humidity test in which painted test panels are subjected to 95 % humidity at 40 °C for 240 hours. This test is also used for water-based, effect-pigmented products. The stability of the effect-pigmented paints is evaluated assessing the following characteristics:

- swelling
- blister formation
- adhesion to substrate
- loss of gloss
- changes in DOI

Practical experience shows that the humidity stability of weather-stabilized, special effect pigments in current automotive paints is good to very good. The performance of the individual pigments is, however, dependent on the paint system in question. The performance can be optimized by a modification of the surface treatment.

The humidity test is a reliable way of assessing the stability of effect pigments. The individual parameters accurately reflect the state of the complete paint system before and after testing.

Photoactivity

Photoactivity (see Chapter 3.5) is an important criterion in pigment evaluation, in particular with regard to use out of doors. This is particularly true of pigments containing titanium dioxide. Iron oxide containing pigments are not photoactive. Practice has shown that effect pigments containing titanium dioxide are intrinsically photoactive. The degree is dependent on their form – rutile or anatase – and the type of surface treatment. Rutile pigments show better stability than anatase. Surface treatment gives a significant improvement in stability.

Chemical stability

Effect pigments are stable against attack from diluted acids and bases. Furthermore, there is no evidence that paint components such as additives etc. have a negative effect on them either.

Thermal stability

Non-stabilized effect pigments in the powder form are stabile up to around 900 °C. Thus, the paint-curing process has no effect on them. In general, pigments that have been treated for weather resistance are stabile to 200 °C.

Stability of the visual effect

Florida test

For materials such as exterior paints and automotive paints etc., in order to guarantee the satisfactory weather stability of the complete paint system, it is essential to subject prototypes of effect paints to natural weather conditions and assess their performance. In the automotive industry, samples are put on exposure in Florida. A pigmented paint system must show a satisfactory result after at least two years exposure (Figure 3.11, page 124).

Gloss and DOI of paint surfaces

The quality of the paint surface is judged by assessing the gloss and hold out (DOI) of the clear coat. This is governed by the orientation of the pigment platelets in the base coat. In general, good parallel orientation gives good gloss and DOI results (Figure 3.12, page 125).

Circulation stability

In most cases, effect pigments are stabile in circulating systems [22, 105]. Since the particles do not distort no color change is observed after a period in a circulating system. Substrate-based effect pigments are relatively hard, thus a suitably hard material such as tungsten carbide should be used for the nozzles and valves.

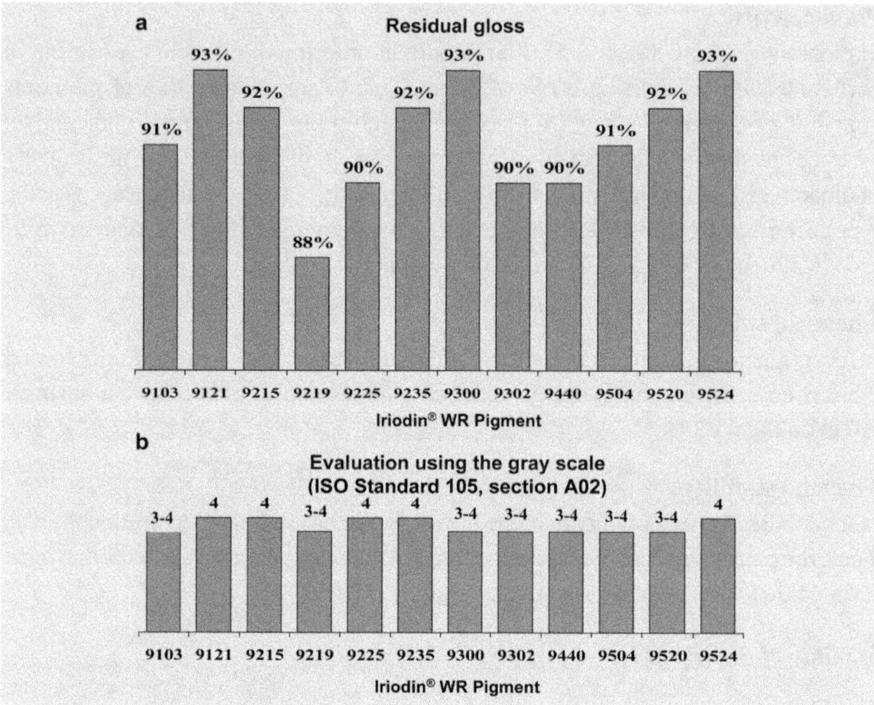

Figure 3.11: Florida test results for an automotive basecoat/clear coat effect coating. Test conditions: 24 months in South Florida, 5°, black box; pigmentation: 4 % "Iriodin" WR-pigment relating to the ready-to-spray formulation. (a) evaluation according to residual gloss, 20°; (b) evaluation using the gray scale (ISO-Standard 105, section A02).

3.11 Examples of use

Here are just some of the many uses of effect pigments in paint systems.

Automotive coatings and automotive repair paints [2, 106]

Effect paints have been used in vehicle painting for 30 years. At the beginning they were employed to give metallic effect paints. At the time, single-coat paints were still being used. They had relatively poor weather resistance. 25 years ago, the first base coat-clear coat effect paints appeared on the market on such legendary cars as the NSU RO 80 [25]. The next step was to test the pearlescent effect pigments used in the cosmetic industry. After initial appearance problems, the first pearl luster effect paints were soon introduced in automotive coatings. Today, depending on the car model, effect paints represent up to 80 % of OEM automotive paints (OEM = original equipment manufacturing) [25]. Of these, some 40 to 50 % of the cars are painted with so-called "mica paints". Base coat-clear application is by far the most

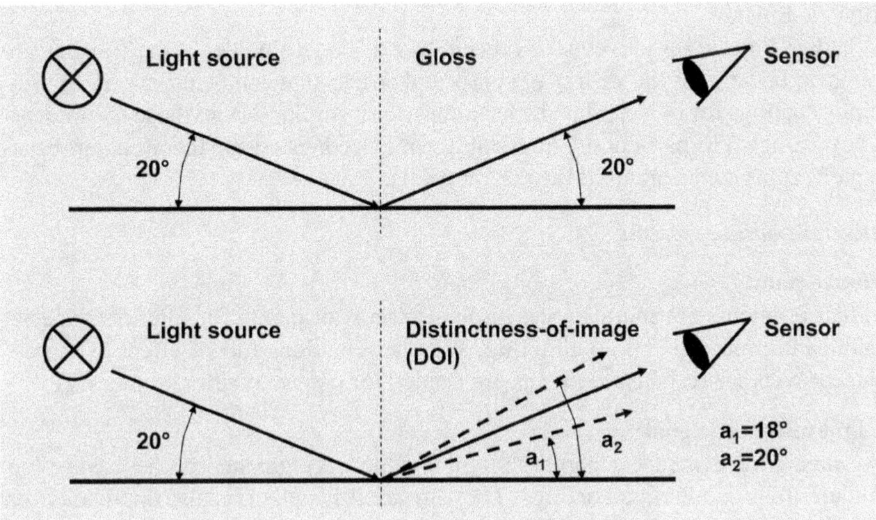

Figure 3.12: Gloss and DOI (Distinctness-of-image) of paint surfaces.

frequently used application method. Tri-coats are the exception. These colors are popular, because of the brilliant and strong effects that can be produced. A consequence of the wide use of these pigments in automotive coatings is their introduction into repair paints. In both cases, weather-resistant, surface treated effect pigments are used (see Chapter 3.5).

Industrial paints [8]

There are diverse uses for effect pigment paints in the industrial sector. Here are some examples:

Can coating

Cans are often coated at the first stage of production, i.e. the complete metal plate, is treated. The coating process is somewhat like printing. This similarity is apparent in the pigment/binder ratio of the products used. These factors must be taken into consideration when designing effect pigmentations for use in can coatings.

Furniture finishing

In recent years, effect pigments have been increasingly used in kitchen furniture, where the silky effect, which cannot be produced with conventional pigments. They are increasingly attracting the attention of architects and designers. Effect paints are not only interesting for the decoration of furniture; they are also being used more and more for interior fittings. This is interesting, because it requires the development of new concepts in effect paints.

Bicycle finishes
Whether for use on bicycles or other two-wheeled vehicles, effect finishes are popular, because of the variety of colors and effects that can be produced. In principle, coatings for bicycles can be formulated in a similar way to those for automotive coatings. The particularly high stability requirements mean that weather-resistant effect pigments are used in such materials.

Diverse coating systems

House paints
Effect pigments are mainly used in dispersion wall paints for interiors. Special application methods allow glittering, aesthetic creations full of effect to be produced. Weather-resistant pigments are needed for use on exterior surfaces.

Light anti-static coatings
As already described in Chapter 2, light conductive pigments can be used to formulate light conductive coatings. They are used for electrostatic application on plastics and to create anti-static flooring [107, 108].

Special uses
A special area is the marking of seeds with a pearl luster coating. This is carried out for a variety of reasons:

- slow down germination
- stop birds eating the seeds
- water storage due to a reduction in a tendency to diffusion
- differentiate seed types by color
- differentiation to competition

Such coatings are applied by a special tumbling process.

References

[1] DIN 55943:2001-10, Colorants (terms)
[2] *Richter, G.*, "Trends bei der Farb-Applikation und Effekt-Lackierung", JOT, **3** (2005) 38
[3] *Finkenzeller, M.* in "Car Colour Management", PPCJ, 11 (2005) 22
[4] "Die Köpfe hinter dem Design", Automobil-Produktion, **11** (2005) 58
[5] *Dworschak, M.*, "Trendforschung: Das Farbkartell", Der Spiegel, **25** (2006) 126
[6] *Maile, F. J., Sharrock, S., Reynders, P.*, "Breathing Life into Cars", PPCJ, **2** (2004) 15
[7] *Maile, F. J., Sharrock, S., Reynders, P.*, "Pigments on the Move", PPCJ, **4** (2004) 24
[8] *Huber, A.*, "Use of Pearl Lustre Pigments in Industrial Coatings", The Coatings Agenda Europe, 2003
[9] *Welsch, N., Liebmann, C. C.*, "Farben – Natur, Technik, Kunst", Spektrum Akademischer Verlag Berlin, 2003

[10] *Wheeler, I. R.*, "Metallic Pigments in Polymers", Rapra Technology Ltd., 1999
[11] *Givargizov, E. I.*, "Highly Anisotropic Crystals", Springer Netherland, 1986
[12] *Fettis, G.*, "Automotive Paints and Coatings", Wiley-VCH, 1995
[13] *Pfaff, G., Reynders, P.*, "Angle-Dependent Optical Effects Deriving from Submicron Structures of Films and Pigments", Chem. Rev., **99** (1999) 1963
[14] "Industrial Inorganic Pigments", ed. G. Buxbaum, G. Pfaff, 3rd edition, Wiley-VCH, 2005
[15] *Klein, G. A.*, "Farbenphysik für industrielle Anwendungen", Springer Verlag, Berlin, 2004
[16] *Schmidt, C., Friz, M.*, "Optical Physics of Synthetic Interference Pigments", Kontakte (Darmstadt) (2) (1992) 15
[17] *Pfaff, G.*, "Industrial Inorganic Pigments", 3rd ed., Wiley-VCH, 2005, 230
[18] *Boxall, J., von Fraunhofer, J. A.*, "Paint Formulation", Industrial Press Inc., New York, 1981
[19] *Müller, B., Poth, U.*, "Coatings Formulation", Vincentz Network, Hannover, 2006
[20] *Mollet, H., Grubenmann, A.*, "Formulierungstechnik", Wiley-VCH, 1999
[21] *Flick, E. W.*, "Water-Based Paint Formulations", Vol. 4, Noyes Publications, Westwood/USA, 1997
[22] *Goldschmidt, A., Streitberger, H.-J.*, BASF-Handbook Basics of Coatings Technology", Vincentz Network, 2nd ed. Hannover, 2007
[23] *Funke, W.*, "Problems and Progress in Organic Coatings Science and Technology", Progr. Org. Coat., **31** (1/1997) 5
[24] *Laible, R.*, "Umweltfreundliche Lackiersysteme für die industrielle Lackierung", Expert-Verlag, 1989
[25] *Hoffmann, P., Duschek, W.*, "Neue Effektpigmente – Darstellung der Effektarten und ihre Auswirkungen auf das Lackierergebnis", Berichtsband DFG, **41** (1999) 123
[26] *Grefenstein, A., Kaymak, K.*, "Exterior Vehicle Body Parts: Films instead of Painting", Kunststoffe international, **8** (2003) 37
[27] *Müller, A.*, "Einfärben von Kunststoffen", Hanser Verlag, 2002
[28] *Maile, F. J., Reynders, P.*, "Effektpigmente in der Lackanwendung", Coating (4/2001) 137
[29] *Stegen, H.*, "Gewusst wie – Kunststoffe richtig lackieren", Metalloberfläche, **57** (2003) 3
[30] *Lommatzsch, U.*, "Vorbehandeln mit Plasma bei Atmosphärendruck", JOT, **6** (2005) 58
[31] Ken-React Reference Manual, Kenrich Petrochemicals Inc., Bayonne, New Jersey, USA, 2nd revised ed., 1993
[32] *Dössel, K. F.*, "On the Way to Zero Emission Automotive Coatings", Int. Conf. Carbody Painting, Berlin, 2005
[33] *Kaluza, U.*, "Physical/chemical fundamentals of pigment processing for paints and printing inks", Edition Lack und Chemie, Elvira Moeller GmbH, 1981
[34] *Oyarzún, J. M.*, "Pigment Processing", Vincentz Verlag, Hannover, 2000
[35] *Patterson, D.*, "Pigments – An Introduction to Their Physical Chemistry", Elsevier Publishing, Amsterdam, 1967
[36] *Turner, G. P. A.*, "Introduction to Paint Chemistry and Principles of Paint Technology", 2nd ed., Chapman and Hall, London, 1980
[37] *Gesenhues, U.*, "Rheology, Sedimentation and Filtration of TiO_2 Suspensions", Chem. Eng. Technol., **26** (2003) 25
[38] *Kresse, P.*, Conf. Proceed., p. 205, IX. FATIPEC-Congress, Brüssel, 1968
[39] *Stech, M., Reynders, P.*, "Emulsions, Foams, and Thin Films", Marcel Dekker, New York, 2000, 447

[40] DIN ISO 9277:1995, "Bestimmung der spezifischen Oberfläche von Feststoffen durch Gasadsorption nach dem BET-Verfahren"

[41] *Hildenbrand, V. D., Doyle, S., Fuess, H., Pfaff, G., Reynders, P.*, "Crystallisation of Thin Anatase Coatings on Muscovite", Thin Solid Films, **304** (1997) 204

[42] *Stech, M., Reynders, P., Rödel, J.*, "Constrained Film Sintering of Nanocrystalline TiO_2", J. Am. Ceram. Soc., **83** (2000) 1889

[43] *Brock, T., Groteklaes, M., Mischke, P.*, European Coatings Handbook, Vincentz Verlag, Hannover, 2000

[44] *Ruf, J.*, "Organischer Metallschutz", Vincentz Verlag, Hannover, 1993

[45] *Gehm, L.*, "Rheologie", Vincentz Verlag, Hannover, 1998

[46] *Reynders, P., Maile, F. J.*, "Sedimentation-Behaviour of Substrate-Based Effect Pigments", XXVII. FATIPEC-Congress, Aix-en-Provence, France, 2004

[47] *Winkler, J.*, "Titan Dioxide", Vincentz Verlag, Hannover, 2003

[48] *Gerthsen, C., Kneser, M., Vogel, H.*, "Physik", 15th ed., Springer Verlag, Berlin, 1986

[49] *Tadros, T. F.*, "Solid/Liquid Dispersions", Academic Press, London, 1987

[50] *Maile, F. J., Reynders, P., Fritsche, K., Pironti, G.-F.*, "Ergänzende Untersuchungen zur Charakterisierung von Effektpigment-Sedimenten", APi-Tagung (GDCh), Eisenach, 2005

[51] *van Olphen, H.*, "An Introduction to Clay Colloid Chemistry", Interscience (Wiley), New York, 1963

[52] *Jasmund, K., Lagaly, G.*, "Tonminerale und Tone", Steinkopff Verlag, Darmstadt, 1993

[53] *Hofmann, U.*, "On the structure of sediments formed by layered silicates", Chemie, **55** (1942) 283

[54] *Weiss, A., Frank, R.*, "Über den Bau der Gerüste in thixotropen Gelen", Z. Naturforsch., **16B** (1961) 141

[55] *Fitzsimmons, R. F., Posner, A. M., Quirk, J. P.*, "Electron microscopic and kinetic study of the flocculation of calcium montmorillonite", Israel J. Chem., **8** (1970) 301

[56] *O'Brien, N. R.*, "Fabric of kaolinite and illite floccules", Clay. Clay. Miner., **19** (1971) 353

[57] *Zou, J., Pierre, A. C.*, "Scanning electron microscopy observations of card-house structures in montmorillonite gels", J. Mater. Sci. Lett., **11** (1992) 664

[58] *Horn, R.*, "Surface forces and their action in ceramic materials", J. Am. Ceram. Soc., **73** (1990) 1117

[59] *Bhatty, J. I., Davies, L., Dollimore, D., Zahedi, A. H.*, "The Use of Hindered Settling Data", Surf. Technol., **15** (1982) 323

[60] *Dulog, L., Hilt, M.*, "Die Stabilität von wässrigen Pigment- und Polyvinylacetatdispersionen", Farbe + Lack, **95** (1989) 793

[61] *Vincent, B.*, "The Effect of Adsorbed Polymers on Dispersion Stability", Adv. Colloid Interfac., **4** (1974) 193

[62] *Kresse, P.*, "Eine chromatographische Methode zur Bestimmung der Flokkulation von Pigmenten", Farbe + Lack, **76** (1970) 351

[63] *Overbeek, J. T. G.*, J. Mater. Educ., **7** (1985) 395

[64] *Patton, T. C.*, J. Paint Technol., **38** (1966) 387

[65] *Buscall, R.*, Colloid Surface, **43** (1990) 33

[66] *Mezger, T. G.*, "The Rheology Handbook", 2nd ed., Vincentz Network Hannover, 2006

[67] *Goddard, E. D.*, "Polymer/Surfactant Interaction in Applied Systems", J. Soc. Cosmet. Chem., **22** (1999) 181

[68] *Legrand, P.*, "Stabilizing Properties of Polymers", ECJ, 3/2005, 90

[69] *Karsa, D. R., Davies, W. D.*, "Waterborne Coatings and Additives", Royal Society of Chemistry, London, 1995

[70] *Pashley, R. M., Karaman, M. E.*, "Applied Colloid and Surface Chemistry", John Wiley & Sons, Chichester, England, 2004

[71] *Henrich, V. E., Cox, P. A.*, "The Surface Science of Metal Oxides", Cambridge University Press, 1994

[72] *Butt, H.-J., Graf, K., Kappl, M.*, "Physics and Chemistry of Interfaces", Wiley-VCH, 2003

[73] *Croll, S.*, "DLVO theory applied to TiO_2 pigments and other materials in latex paints", Progr. Org. Coat., **44** (2/2002) 131

[74] *Thometzek, P., Ludwig, A., Karbach, A., Köhler, K.*, "Effects of morphology and surface treatment of inorganic pigments on waterborne coating properties", Progr. Org. Coat., **36** (4/1999) 201

[75] *Bosch, W., Schlesing, W., Buhk, M.*, "Pigments effect mechanical coating properties", ECJ, 10/2001, 60

[76] *Perera, D. Y.*, "Review: Effect of pigmentation on organic coating properties", Progr. Org. Coat., **50** (4/2004) 247

[77] *Cabral, A. M., Duarte, R. G., Montemor, M. F., Ferreira, M. G. S.*, "A comparative study on the corrosion resistance of AA2024-T3 substrates pre-treated with different silane solutions", Progr. Org. Coat., **54** (4/2005) 322

[78] *Maile, F. J., Pfaff, G., Reynders, P.*, "Effect Pigments: Past, Present and Future", Progr. Org. Coat., **54** (3/2005) 150

[79] *Siemen, A.*, "Untersuchung der Bronceverteilung bei der Hochrotationszerstäubung von Metalleffektlacken im Hinblick auf die Effektausbildung", Diplomarbeit, Universität Paderborn, 1990

[80] *Domnick, J.*, "Neue Erkenntnisse zur Hochrotationszerstäubung", JOT, **6** (2006) 32

[81] *Ondratschek, D.*, "Forcierung des Wasserlackeinsatzes durch neue Trocknungsverfahren", Institut für Industrielle Fertigung und Fabrikbetrieb, Universität Stuttgart, BWSE 99003, 2000

[82] *Yang, C., Smyrl, W. H., Cussler, E. L.*, "Flake Alignment in Composite Coatings", J. Membrane Sci., **231** (1–2/2004) 1

[83] *Funke, W.*, "Preparation and Properties of Paint Films with Special Morphological Structure", JOCCA, **59** (1976) 389

[84] *Maile, F. J., Reynders, P.*, "Pigmentos de efeito para o novo milena", Vol. 2 (2003) 435, 7° Congresso International de Tintas (ABRAFATI)

[85] *Geilen, S.*, "Korrelation zwischen den optischen Eigenschaften von Effektbeschichtungen und der Orientierung der enthaltenen Effektstoffe", Diplomarbeit, Universität Paderborn, 2002

[86] *Maile, F. J., Rösler, M.*, "Local Gloss and Sparkle caused by Effect Pigments: Psychophysics, Measurements and Simulations", XXVI. FATIPEC-Congress, Dresden, 2002

[87] *Kirchner, E., Njo, L., de Haas, K., Rösler, M., Gabel, P.*, "More than colour", ECJ (11/2006), 46

[88] *Goldschmidt, A.*, Deutsches Wasserlacksymposium des Praxis-Forums, Bad Nauheim, 1992

[89] *Bendel, A.*, "Farbliche Gestaltung mit dem Glanz der Perlen", Der Maler und Lackiermeister, **10** (1993) 894

[90] *Bendel, A.*, "Alte Technik – Neuzeitlich Angewandt", Der Maler und Lackiermeister, **3** (1997)

[91] *Bendel, A.*, "Neue Technik – Glanzvergoldung im Pulververfahren", Der Maler und Lackiermeister, **12** (1999)

[92] *Bendel, A.*, "Lasur-Technik: Stilvolle Gestaltung – Sichere Systemanwendung", Der Maler und Lackiermeister, **7** (2000)

[93] *Slinckx, M., Henry, N., Krebs, A., Uytterhoeven, G.*, "High-solids automotive coatings", Progr. Org. Coat., **38** (3/2000) 163

[94] *van den Haak, H. J. W., Krutzer, L. L. M.*, "Design of Pigment Dispersants for High-Solids Paint Systems", Progr. Org. Coat., **43** (1/2001) 56

[95] *Dören, K., Freitag, W., Stoye, D.*, "Wasserlacke", Verlag TÜV Rheinland, Köln, 1992

[96] *Bates, D. A.*, "The Science of Powder Coatings", Vol. I, Gardiner House, London, 1990

[97] *von Keudell, L.*, "Strömungsverhältnisse in der Kabine – Optimaler Lufthaushalt sichert effiziente Pulverapplikation", JOT, **6** (2006) 26

[98] *Rembert, A.*, "Vollautomatische Lackieranlage bei John Deere – Landmaschinen in Automobil-Qualität", JOT, **6** (2006) 20

[99] *Bopp, M.-L.*, "Infrarot-Booster beschleunigt die Fertigung – Pulverlack effizienter härten", JOT, **6** (2006) 24

[100] *Rolles, R., Williams jr., J. E., Kondis, T. J.*, US 4,003,872

[101] *Hosokawa, M., Morimoto, M., Iwato, M.*, JP 88-64418

[102] *Kieser, M.*, EP 0 285 977

[103] *Dorfner, K., Ohngemach, J.*, Paint India, (1/1987) 45

[104] *Maisch, R., Stahlecker, O., Kieser, M.*, "Mica pigments in solvent-free coatings systems", ECJ, (9/1994) 582

[105] *Betten, P.*, "Kolloidchemische Untersuchung zur Optimierung von Autoserienlacken", Dissertation, Universität Kiel, 2003

[106] *Rink, H.-P., Mayer, B.*, "Water-based coatings for automotive refinishing", Progr., Org. Coat., **34** (1-4/1997) 175

[107] *Brückner, H.-D., Glausch, R., Maisch, R.*, "Neuartige helle, leitfähige Pigmente auf Glimmer/Metalloxid-Basis", Farbe + Lack, **96** (1990) 411

[108] *Glausch, R., Pfaff, G., Maisch, R.*, "New Results with Light Colored Conductive Pigments", XXIst FATIPEC Congr., Amsterdam, Bd. 2, 1992, 33

[109] Reprinted from Progress in Organic Coatings, Vol. 44, *Croll, S.*, "DLVO theory applied to TiO_2 pigments and other materials in latex paints", page 131, (2/2002), with permission from Elsevier

[110] Reprinted from Progress in Organic Coatings, Vol. 54, *Maile, F. J., Pfaff, G., Reynders, P.*, "Effect Pigments: Past, Present and Future", page 150, (3/2005) with permission from Elsevier

Figure 1: Images concerning the DIN 55945 (1999) definition of an effect paint – (a) angle dependency of the color (i.e. the interference color is visible in the gloss angle and invisible outside of the angle), (b) multi-color effect or dependency of the color on the observation angle (i.e. with change of the observation angle, the interference color also changes, so-called goniochromatic properties [13]), (c) structure/texture (see Chapter 3.1).

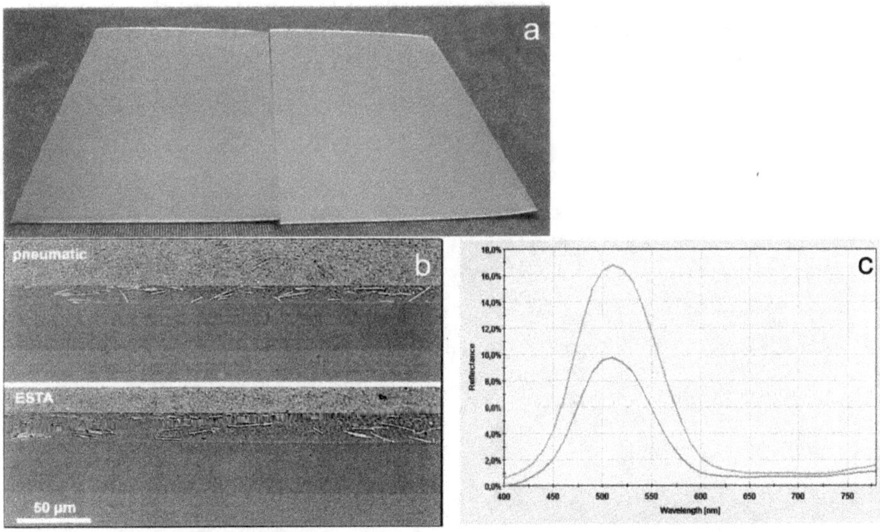

Figure 2: Photographs of steel panels painted in a car-manufacturing process. The basecoats used contain 2.5 % of an industrially produced titanium dioxide/mica effect pigment (rutile) and were applied in different ways: (a) with high-rotation atomization (ESTA) (left) and Spraymate (right), (b) cross sections (pneumatic and electrostatic), (c) remission curves (geometry: 45°/120°) of the two panels (upper curve: pneumatic, lower curve: ESTA) [110] (see Chapter 3.6).

Figure 3: Bright field microscopic image of an actual automotive color (styling with inorganic/organic pigments), two-layer coating, Audi color Aura Blue (code Z7R) with TiO_2/corundum effect pigments ("Xirallic" T60-10 SW Crystal Silver, see colored, platelet-shaped particles) (see Chapter 3.8).

Figure 4: Pearl lustrous polyester buttons (see Chapter 4.2.1).

Figure 5: Injection molded parts colored with effect pigments (see Chapter 4.3).

Figure 6: Thermoformed, extruded polymethyl methacrylate sheet colored with a heat-reflecting interference pigment (see Chapter 4.3 und 4.3.1).

Figure 7: Marbled pearl lustrous injection molded parts (see Chapter 4.3.3).

Figure 8: Example of a screen printed product manufactured with a green textile printing ink containing the effect pigment "Miraval" Magic White (based on borosilicate flakes) on a black fabric (see Chapter 5.2.1).

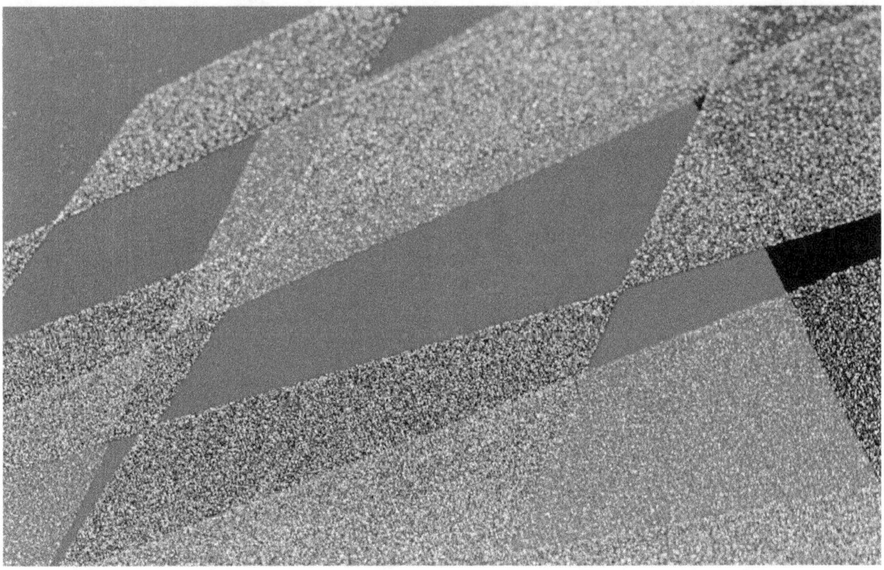

Figure 9: Close-up view of the effect pigments "Colorstream" Autumn Mystery (based on SiO_2-flakes) and "Iriodin" 7235 Ultra Green (based on mica) screen printed self-adhesive label (see Chapter 5.3.3).

Figure 10: Comparison of an iron oxide mica pigment in form of "Iriodin" Pearlets (left) and as pure pigment powder (right) (see Chapter 5.3.6).

4 Special effect pigments in plastics

Dr. Manfred Kieser

In Chapter 2 the physical and chemical requirements are described for effect pigments and in particular for pearl luster pigments and for an ideal pearl effect. This chapter describes more closely the special features in the use of luster pigments in plastic applications.

In the following the term pearl luster pigment is applied for the used pigments. This term stands for all transparent platelet-shaped effect pigments including the transparent interference pigments.

4.1 General information on the application of pearl luster pigments in plastics

The three most important conditions, which must be observed to obtain the optimum effect from pearl luster pigments in plastics, are:

- transparency of the plastic matrix
- complete dispersion of the pigment platelets
- parallel orientation of the pigment platelets in the plastic.

In addition, specific properties of the plastic and/or processing can still affect the results.

4.1.1 Transparency of the plastic matrix

Chapter 2.1 describes how the luster of platelet-shaped pigments is determined by the regular reflection of light upon thin, highly refracting layers. Anything that disrupts or prevents this regular reflection of light from the pigment platelets reduces or destroys the pearl luster. Light scattering substances such as pigments, filling materials, crystallites in the plastics, as well as the disperse phases in a multi-phase plastic all have a negative influence. To obtain highly lustrous effects, one must exclude or avoid using such materials.

Despite these limitations interesting luster effects can be achieved with slightly opaque and moderately filled plastics. In some special cases it can be recommendable to use a certain amount of hiding pigments or fillers to achieve a particular effect (see Chapter 4.3.7).

4.1.2 Dispersion of the pearl luster pigments

A disordered mass of pearl luster pigments exhibits practically no luster at all. At the most, a slight glittering from individual platelets can be observed. As with conventional pigments, the pigment particles must be completely dispersed so that they can be orientated in a parallel manner (see Chapter 4.1.3).

Every dispersion process consists of four different steps, which take place partly one after the other as well as simultaneously. Thus for a singular pigment agglomerate the following steps take place:

- wetting
- dispersing
- distributing
- stabilizing

The faultless implementation of these steps is necessary to ensure a successful and economical pigmenting of the end product. Before a pigment agglomerate can be separated by shearing forces, it must be wetted-out by a liquid phase. Simply moistening the pigments and then shearing them is not sufficient. They must also be adequately and evenly distributed throughout the entire material and, finally, stabilized. The difficulties in the coloring of plastics with pearl luster pigments generally lie in the first and third step, in wetting and distributing.

The shearing forces needed for dispersing the agglomerates, second step, are usually present in conventional thermoplastic processing methods. Sometimes, the forces are so great that damage to the pigment platelets might occur. It is important to ensure that the pigment platelets themselves are not sheared unnecessarily to avoid any damage. It is recommended to process pearl luster pigments at the upper end of the plastic manufacturers recommended temperature range.

The stabilizing of pearl luster pigments, especially in thermoplastics, is generally not a problem. During the cooling phase the viscosity rises rapidly, thus freezing the dispersion state. Hereby re-agglomeration and sedimentation cannot take place. Wetting the pearl luster pigments is the most important step for the dispersion process; this is true not only for application in plastics. The process of wetting can be influenced by thermodynamic and by kinetic factors, although the choice of thermodynamic factors is limited (for example lowering of the surface tension in the wetting phase). Both the pigment surface and the plastic are usually defined for a certain application and therefore modifications are limited.

The possibilities with kinetic factors are much greater. For example, lower viscosity of the wetting phase, good mixing of pigments and liquid phase. These can be varied either by the choice of the processing temperature or the melt viscosity of the selected plastic (higher melt volume rate for the same chemical structure), or with a pre-wetting of the pigment using a lower molecular weight polymeric compatible substance (paraffin oil, plasticizers, etc.).

Despite this, a rapid wetting of the pigments is not easy to do with all plastics. Careful working and care is required for the pearl luster pigmenting of plastics.

The homogeneous distribution of the dispersed pigment platelets throughout the plastic is a very important step, particularly for thermoplastics. Non-homogeneous pearl pigment distribution will be seen as streaks, stripes and color shadows in the final product.

The mixing efficiency of standard processing machines does not always meet expectations. Frequently, it is necessary to optimize materials, machines, as well as processing parameters for each individual application. The distribution process is generally improved using a lower viscosity, turbulent flow, and longer processing times.

4.1.3 Orientation of the pigment platelets

In order to demonstrate pearl luster, the pigments must be aligned parallel to each other and, normally, parallel to the object surface. This occurs in almost all applications through flow movements in the corresponding matrix. In casting resins this happens while the casting process (see Chapters 4.2.1 and 4.2.3). The processing of thermoplastics generally proceeds through a liquid phase, and is therefore also combined with flowing processes. Additional steps to align the pigments are usually not necessary.

Irregular flowing processes in the material, known as flow lines and weld lines in the plastics object, usually lead to undesirable disruption of the pigment orientation, which is disturbingly seen in the finished object. Pearl luster pigments themselves are not the cause of flow lines and weld lines, these are already present and sometimes even visible, but nevertheless can be detected due to varied mechanical properties in the material.

4.1.4 Applications for pearl luster pigments

The large field of plastics can be broken down into various criteria, for example, monomers, polymerization reactions, manufacturing processes or many other additional parameters.

When discussing the application of pearl luster pigments in plastics, it is most practical to separate them into two distinct fields; thermoplastics and duromers. Duromers are occasionally called duroplasts. This is not correct, as duromers are known for their non-thermo plasticity, non-melting, non-soluble characteristics after conversion into an end product.

4.2 Pearl luster pigments in duromers

With regard to the application of pearl luster pigments, duromers demand a further classification. For example, it should be distinguished between duromer materials which are processed in the fluid phase, such as casting and laminating resins, and duromers which are pressed out of a powder material, the so-called molding materials.

Historically, the implementation of pearl luster pigments in polyester casting resins for the manufacturing of buttons was the first application field of industrial importance. Parallel to this, the application of pearl luster pigments in acrylic casting resins, for buttons as well as decorative acrylic sheets, showed some technical importance too.

The use of pearl luster pigments in laminating resins is a relative small application niche. The entire field of press materials in the processing of pearl luster pigments has played a remarkably small role to date. The reasons lie for the most part in the heterogeneity of the press materials and the very slight flow while processing. This is true for both the phenolic resin as well as melamine resin molding materials.

4.2.1 Unsaturated polyester casting resins

The main application area for pearl luster pigments in unsaturated polyester resins is in the manufacturing of buttons, using centrifugal, rod and cup mold casting methods [1, 2].

Pearl luster pigments are particularly suitable for the centrifugal molding process [3–7]. By this process, the polyester material, which already contains the pigment, accelerator and catalyst, is casted in a horizontally arranged rotating steel drum. After a set time period (approx. 30 min) the still flexible sheet is removed, spread out flat and mechanically processed further. The final curing can take place before or during the mechanical manufacturing steps.

Pigment paste application

Special pastes are available to enable an optimal orientation of the pigments during the processing of the liquid resins. They contain additives to improve the distribution of the pigments, as well as additives to improve the pigment orientation.

Years ago, these pastes contained as a pearl luster pigment mainly the toxic basic lead carbonate. Today pastes with bismuth oxychloride were in the market. The tendency of pearl luster pigments to sediment is a great disadvantage concerning their processing. Nevertheless pastes with lead carbonate and bismuth oxychloride were used in unsaturated polyester resins, because with these pigments no high quality products in a dry powder form can be prepared. All other effect pigments were used in this application in powder form but usual in combination with additives to improve distribution and orientation of the pigment platelets.

Pigment concentration

In polyester casting resins, the pearl luster pigments are used in comparatively low concentrations. Normally 0.8 to 1.0 % pigment paste (based on the resin) with 30 to 35 % pigment content is sufficient for a good pearl luster effect. Pigments in powder form are used even in lower concentrations, 0.2 to 0.5 % are usual. For interference pigments, the concentration is slightly higher. This is due to the smaller number of platelets present (percentage per weight) of interference pigments, because interference pigment platelets are somewhat thicker than the silver white pigments of the same particle range type.

To ensure the efficiency of the additives, which improve distribution, it is necessary to use a minimum amount of pigment paste in the liquid polyester resin. The technical term for this amount is the "lowest concentration limit". It varies depending on the unsaturated polyester resin and the quality of the pigment paste. The lowest concentration limit of a good paste should be 0.4 % at the maximum. If the lowest concentration limit is not reached problems in the orientation of the pearl luster pigments in the button sheet will occur. In order to achieve good orientation results, even in such cases, highly concentrated liquid additives are available on the market, which take over the function of the missing fluid phase of the pigment paste. These additives are also used for the processing of powdered pearl luster pigments of the layer substrate principle in polyester casting resins.

The necessary minimum amount of additive is proportional to the polyester resin and not to the pigment amount. These additives also permit the direct application of all powder pearl luster pigments in unsaturated polyester resins. Usually, the amount of such an additive required lies between 0.3 and 0.8 %.

Production of a polyester button sheet

Even modern pigment pastes partly modified by thixotropic additives, which lead to reduced sedimentation properties need to be well stirred before application. Mixing with a simple stirrer is sufficient to implement pigment pastes or pigments. The mixing time should be long enough to enable any bubbles in the polyester dispersion to escape, and to ensure that the pigment is fully dispersed and distributed throughout the resin system. To achieve a good pearl luster effect, the resin viscosity at the processing temperature should, if necessary, be adjusted to 0.8 to 1.0 Pa · s with the reaction component styrene. The gel time of the resin should be between 7 to 10 min (can be varied with the amount of initiator and accelerator used). A standard value for the rotation of the centrifugal drum is a circumferential speed of 3 to 5 m/s; for a drum diameter of 80 to 90 cm this results in a speed of 80 rpm.

The polymerization inhibiting influence of the oxygen in the air, which is noticeable by the stickiness of the air-exposed side of the finished button sheet, can be prevented or eliminated using various techniques. The following possibilities are commonly used:

- Covering the air side of the gelled sheet in the rotational drum with a solution of 2 % liquid paraffin in an aliphatic hydrocarbon solvent.

- Addition of 0.05 to 0.1 % paraffin (mp 42 to 45 °C) in form of a 2 to 5 % solution in styrene. The optimum amount and the paraffin type will depend upon the polyester and the processing temperature.
- Powdering the sticky air side with talcum (usually after removal from the rotational drum).
- Covering the air side with a plastic sheeting (in the drum).

The above-mentioned details are a guide for normal working procedures. For special effects, deviations from these guidelines will sometimes be necessary. The movements (flowing) that take place in the resin during the casting stages are necessary for the good alignment of the pearl luster pigments [8]. Experimental work has clearly shown that pigment platelets align well in flow processes, predominantly due to gravitational pull exerted upon them in the polyester layer [9]. In Figure 4.1, the forces necessary for the alignment and their influence upon the pigment platelets are illustrated.

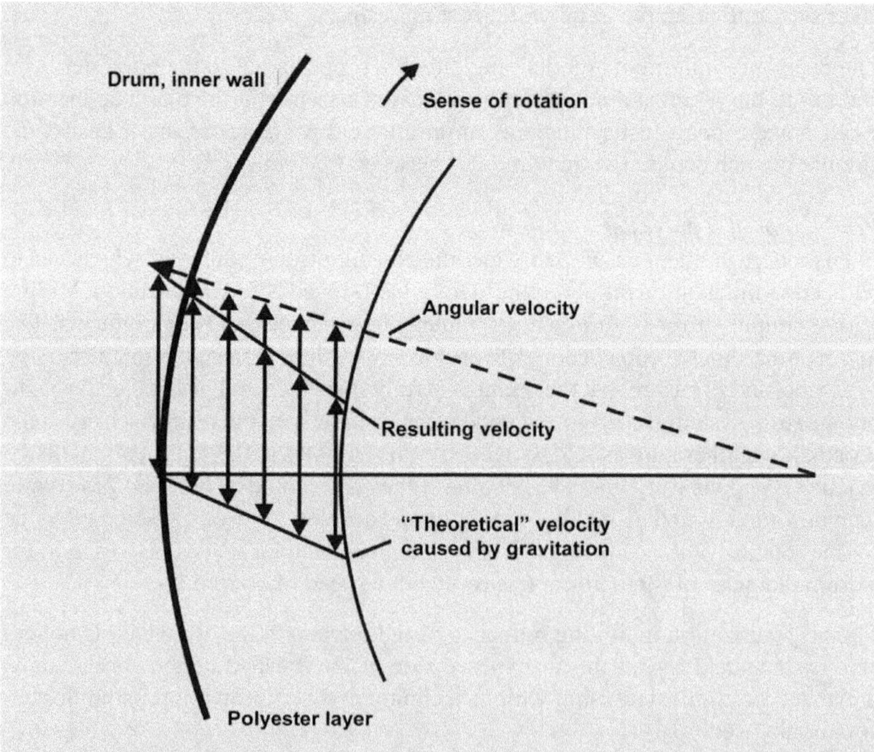

Figure 4.1: Schematic illustration of the flow processes which lead to the orientation of pearl luster pigments in centrifugal castin.

In order to obtain good and uniform pigment orientation, conditions should be such that foreign influences such as air bubbles, dust particles, inhomogeneous polyester resin, etc. should be avoided by correct and clean working methods.

4.2.2 Polymethyl methacrylat casting resins

Pearl lustrous polymethyl methacrylate (PMMA) cast sheets have a broad spectrum of application possibilities. They not only serve as decorative elements for wall designs, but also serve as starting materials for the manufacturing of attractive sanitary furnishings (for example: deep drawn bath tubs, sinks, etc.) as well as for the production of buttons and accessories.

In the past, pearl luster pigments made of basic lead carbonate were mainly used in these applications. In the meantime, non-toxic modern pearl luster pigments have replaced them, which are based upon metal oxide coated natural or synthetic micas, SiO_2 platelets, Al_2O_3 platelets and glass flakes. In certain applications where a very brilliant luster is needed BiOCl pastes are still in use.

There are various working methods in the manufacture of pearl luster PMMA cast resin sheets; the following describe the most common techniques in use.

Application of methyl methacrylate

Firstly the stabilizer must be removed in the usual way from the commercially available, stabilized methyl methacrylate. After adding the appropriate initiator (peroxide or similar radical formers), the monomer is pre-polymerized at 90 to 100 °C to syrup-like consistency (viscosity >5 Pa·s). Before introducing the pigment into this viscous material, the pigment should be pre-moistened with a little of the methyl methacrylate. This step prevents the possibility of entrapping air, and the pigments can be more easily dispersed in the pre-polymerized mass. Experience has shown that a mixture of one part pigment powder and two to three parts monomer produces a good free flowing paste which can be easily stirred into the casting resin. Stirring with a propeller mixer gives a good dispersion of the pearl luster pigments. A homogeneous distribution is achieved after about 10 min, depending upon the size of the batch. According to the desired effect, the necessary hiding power and the thickness of the sheet the pigment concentration is in between 0.15 and 1.0 %. In addition other colorants can be used but they have to be tested for this application (e.g. stability against radicals).

Application of commercial acrylic casting resins

Applications with these commercially available casting resins do not require a separate step for the pre-polymerization. The addition of the pearl luster pigments and the further processing is carried out in the same step as described above for the pre-polymerization of the monomer. Due to the fact that casting resins contain a stabilizer, it may be necessary to select a somewhat higher amount of initiator to "override" the stabilizer.

Forming

Common sheet casting forms are mirror-finish glass plates which are kept separate from one another and sealed at the sides and edges with the use of soft plastic profiles, silicon tubes, etc. The assembly of the sheets takes place with pins, frames or similar tools. To ease dismantling after the resin is polymerized, small amounts of release agents can be added to the casting resin, such as sodium dioctyl sulfosuccinate.

4.2.3 Orientation methods for PMMA cast resin sheets

To achieve an optimum luster effect, the pearl luster pigment platelets must be aligned parallel to the surface. This requires that a certain amount shear flow is present within the material, for the first two methods mentioned below. This is illustrated in Figure 4.2. The shearing turns the platelets as long as the velocity difference $v = v_1 - v_2$ decrease to a minimum. To achieve pigment alignment, the following methods are applied in the practice.

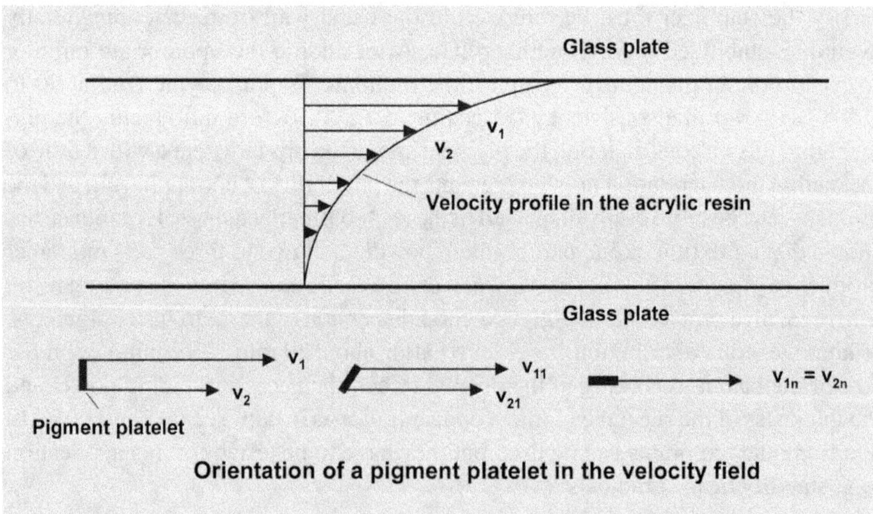

Figure 4.2: Schematic illustration of the flow processes which lead to the orientation of pearl luster pigments in casted acrylic sheets. The velocity vectors v_1 and v_2 influence the pearl luster pigment platelets, thus turning the platelets and creating the new vectors v_{11} and v_{21}. This repeats itself until $v_{1n} = v_{2n}$. Then the pearl luster pigment platelets are oriented parallel to the surface

Orientation by vibration

The molds are set into a longitudinal oscillation using a vibration device set in the plane of the sheet. The sheet molds can be arranged either vertically (for smaller sheets) or horizontally. A warm water or air bath can be used to control temperature; the bath temperature will depend upon the initiator used (usually between 60 and 90 °C). After a polymerization time of about one hour, the resin is usually so viscous that the pigment platelet alignment achieved by vibration remains fixed.

Manual orientation

With this technique the resin is polymerized in the mold up to a point where it still can flow but is viscous enough to sufficiently fix the pigment platelets (the exact reaction time has to be determined and optimized experimentally). At this point, the assembly clips of horizontal glass sheets are released and the upper sheet is gently slid over the lower sheet several times (a few millimeters are sufficient, even for large sheets). The considerable increasing luster indicates the pigment orientation immediately. If this does not happen, then the orientation procedure was carried out too late in the polymerization reaction. After the pigment platelet alignment has taken place, the glass sheets can then be refixed again, until full polymerization of the polymethyl methyacrylate has been completed. Should the achieved luster then decrease, then orientation was carried out too early and should be repeated. This orientation method provides a highly lustrous, uniform shine.

Another method of manual orientation which produces attractive patterns and cloud effects is to move objects (for example, shaped wires, magnet bars or similar items) slowly through the fluid material. The requirements for fixing the alignment are identical as in the above-mentioned cases, particularly regarding the time.

Orientation by sedimentation

If the pearl luster effect is only required on one side, in contrast to the above, a simpler procedure can be implemented. No additional steps are necessary to ensure orientation.

The pigment is stirred into the monomeric methyl methacrylate or in a highly liquid pre-polymer. The mold is sealed tightly after pouring in the suspension into it and the material is polymerized in the horizontal position. Due to the highly liquid consistency, the relatively large and dense pigment platelets quickly sediment and build an even pigmented layer at the bottom of the sheet. This procedure is mainly used when effect pigments were used in combination with other colorants, and produces opaque colored PMMA sheets with an excellent, one-sided pearl luster effect.

Orientation in rotational molding

This procedure, which is in popular use in the Far East, consists of tubes made of glass or metal with a highly polished inner surface, which can be sealed at both ends.

Due to the high evaporation rate of methyl methacrylate, open rotational drums, conventionally used for rotational molding of UP-resins, are not suitable for PMMA as a skin builds up on surface of the PMMA suspension which would lead to the complete disorientation of the pigments. Furthermore the concentration of monomer fumes in the vicinity of the production can cause dangerous pollution levels.

The orientation of the pigment platelets occurs as a result of the same flow processes described for UP button sheets (Chapter 4.2.1). The pigmented PMMA pre-polymer is poured into the drums or tubes, which are then tightly sealed. The thickness of the sheets is proportional to the amount of pre-polymer introduced into the vessel. The polymerization occurs whilst a driven roller bank is rotating the tubes. The rotational speed is adjusted to create a uniform layer within the tube without, in turn, centrifuging the pigments. After the polymerization is finished, the PMMA is removed from the mold, cut apart, and flattened out into a sheet under heating. The material should not be fully cross-linked, but should still remain somewhat thermoplastic.

General notes for PMMA sheet manufacturing

The following points should be observed during manufacturing to achieve an optimal sheet:

Very clean working conditions

Cleanliness should be observed throughout the entire manufacturing process. The molds and tools must be cleaned before assembly. The assembly should be carried out in a dust-free atmosphere. Even during the production of the PMMA solution, utmost care should be taken that no impurities (for example dust, dirt, dried prepolymer, etc.) are introduced into the dispersion.

Complete degassing of the PMMA dispersion

It is essential to completely eliminate any air bubbles in the dispersion in order to produce bubble-free PMMA sheets, usually by gentle stirring slightly below or at the polymerization temperature. The degassing is finished when no further bubbles are to be seen in the warmed dispersion. The degassing time depends upon the temperature, viscosity, and the batch size of the dispersion. Degassing is especially successful under vacuum, although monomer loss must be accounted in this case.

Uniform, smooth pouring into the mold

The mold should be heated to the same temperature as the dispersion in order to pour the solution without problems into the mold. This is particularly important, because mistakes made at this stage of pouring (e.g. material hardening in cold zones) cannot be corrected. The use of a broad funnel has proven to be very useful while pouring. It is also important to take care that no air is entrapped during the pouring of the resin dispersion.

Naturally, this summary cannot encompass all available techniques. As the manufacturing of pearl lustrous PMMA sheets still falls within the category handicrafts, and the results of which are very much dependent upon the experience, knowledge and the craftsmanship of the workers involved, as well as upon the creativity of the designers and manufacturers.

4.2.4 Laminating resins

Laminating resins are processed both with and without pressure. Polyester and epoxy resins belong to the group of resins mostly processed without pressure; whereas phenol and melamine resins are processed under very high pressure.

For all of these applications pearl luster pigments play a very minor role, because the implementation in these highly filled products rarely leads to satisfactory effects. There are, however, a few examples in which a reasonable pearl luster effect can be achieved.

Unsaturated polyester and epoxy resins

The application of polyester and epoxy laminated resins are done almost exclusively in combination with fiber-shaped reinforcing materials such as woven fabrics, non-woven materials and rovings out of glass fibers, carbon fibers, or polymeric fibers. Naturally, the parallel alignment of the pigments and thus a notable luster effect is impossible within such highly filled combination materials, due to the fiber content. Pearl luster pigments are therefore not used in the fiber-reinforced layers of the laminate, but rather more often in the transparent over-coatings.

These over-coatings are also called gel coats. Since these relatively thick coatings (approx. 0.5 mm) scarcely shrink during hardening, a good alignment of the pigments in the final product is only possible when the viscosity rapidly rises after application so that the original pigment orientation is fixed.

The application of the gel coat finish is usually by brushing, spatula or by spraying. Often traces of brush or spatula marks are visible in the luster effect. This is frequently and advantageously used as a design element. Applying the gel coat as a two-coat system has its advantages, in that the second coat can be a highly pigmented thin pearl lustrous layer, and the first as a clear non-pigmented layer. This technique can be found in applications for the manufacturing of luxurious sanitary furnishings and elements for interior design.

Phenol and melamine resins

These resins are predominantly used for the production of laminated sheets, but also for pressing materials. The application of pearl luster pigments in pressing materials will be discussed in the next chapter; this chapter handles the use in laminated sheets.

Due to their own dark color, phenolic resins only permit dark and covering shades. For this reason phenolic resins are rarely used in the outer decorative layers of laminated sheets, and thus, unimportant for pearl luster pigments.

Melamine resins are the material of preference for decorative laminated sheets. They are commonly found as the laminating resins in the colored layers and the overlay papers of laminated sheets. Sheets of this type have a broad application spectrum such as furniture, working surfaces (tables, desks) and wall covering panels, to name a few. Unfortunately, pearl luster pigments in melamine resin sheets are not free of problems, which are underlined by the very small market percentage at this time.

In this application, the pearl luster pigment is not directly introduced into the resin, but printed on the so-called decor paper by gravure printing (mode of operation see Chapter 5.3.1). Decor papers printed with pearl luster pigments demonstrate a good luster after printing and before pressing. After pressing combined with an overlay paper, there is not much that remains of the pearl luster effect. The cause of this loss of luster lies in the optical properties (primarily opacity) of the cured overlay-paper. It has the transparency of parchment paper, which allows the printed colorful ornaments of the decor paper to appeal when spread out upon it, but works as a diffuser (screen) upon direct reflected rays of light. Pearl luster and metallic shining effects cannot be achieved with this construction. The pearl luster effects can be achieved either without an overlay, or with an overlay that is not completely closed, for example an open net or fence structure. In most applications the overlay is necessary for a durable surface quality, hence the luster effects are not always achievable under this criteria.

Pearl luster pigments consisting of titanium dioxide and flaky substrates demonstrate a remarkable phenomenon under exposure to light in these applications. The reason for this is the titanium dioxide, which, when present in an unstabilized form, undergoes a partial reduction in the absence of oxygen, the presence of formaldehyde and under the influence of light. This reaction can be observed as a graying effect [10]. It can be avoided using special stabilized pigments. Under the influence of oxygen this graying reaction, which can be seen also in many other applications, is reversible and normally regenerates very fast after the irradiation with light. But in those melamine sheets the graying regenerates very slowly, it is therefore probable, that traces of formaldehyde intensifies and stabilizes this effect.

4.2.5 Compression molding compounds

In compression molding materials the phenolic and melamine resins play a dominant role. Both materials are relatively pressure and temperature resistant plastics. Due to their natural color, phenolic resins produce only dark colorings, whilst melamine resins can be slightly light to dark color shades. There are relatively few applications with pearl luster pigments in this segment, because the achievable effects in the compressed end product are usually not satisfying.

One reason is the extremely low transparency of highly filled compression molding material. Another reason is the limited orientation possibility for pearl luster pigments in this matrix, due to the restricted flow processes in the compressing machine.

In some structured decorative designs, limited flow is possible to be seen. However, these effects cannot be compared to the normal luster effects and, unfortunately, are not predictable and often not reproducible.

4.3 Pearl luster pigments in thermoplastics

Pearl luster pigments of the layer substrate principle type are very suitable for the pigmentation of practically all thermoplastics. Nowadays they are "classical" pigments for use in plastics.

The application of pearl luster pigmented thermoplastics is predominantly in the packaging field (bottles, lids, cans, inlays etc.) as well as household articles, toys, profiles, sheeting (Figure 4.3), and others. There are applications like building materials such as guttering, heat reflecting sheets for roof light domes, multi wall sheets for winter gardens and roofing which are all state-of-the-art. Even in the interior decoration of vehicles mass-pigmented parts e.g. decorative rings, facings and door coverings are in use.

The majority of modern, layer substrate pigments can be used in nearly all thermoplastics. Most of them conform to the governmental and other regulations which are applicable to materials used in the manufacture of children's toys and food packaging products. But nevertheless the national regulations have to be considered. The pigment producer will give on demand detailed information for this purpose.

Figure 4.3: Thermoformed polypropylene film colored with a golden effect pigment

4.3.1 Effect designing

As already described in Chapter 4.1.1, the luster effect achievable depends not only upon the type of pigment, but most importantly it depends on the transparency of the medium to be colored. For this reason it is important to note that thermoplastics cover the entire field from highly transparent materials over slightly opaque products to strongly opaque, through to non-transparent opaque materials. Therefore, using the appropriate combinations of pearl luster pigments, plastic type and, if necessary, further colorants, almost any imaginable design effects can be created.

Highly lustrous effects will always be obtained when the plastic is clear and transparent. Such transparent polymers in form of sheets are used amongst others for heat protection glasses in roof light domes.

But when coloring, less transparent or even opaque plastics, the luster and iridescence of the plastic part will be reduced. The best effects can be achieved with highly transparent, brilliant plastics, such as polystyrene (PS) or polymethyl methylacrylate (PMMA). But even less transparent and less brilliant plastics, such as polypropylene (PP), low density polyethylene (PE-LD), as well as transparent, modified polystyrenes (ABS) and transparent polyamide (PA) can provide satisfactory pearl luster effects.

Even with opaque plastics, such as high density polyethylene (PE-HD), or opaque ABS and PA types, luster effects can be achieved, although in these cases the "intensity" of the pearlescent colorings will be considerably diminished.

Acceptable luster effects obtained with filled plastics (talcum, glass fiber, chalk, or something similar) are generally speaking, an exception to the role.

While adding conventional colorants, it is important to ensure that only dyestuffs or transparent pigments are used and not strongly opacifying materials because this will reduce or destroy the regular reflection of the light in the medium.

4.3.2 Pigment selection and pigment concentration

Effect pigments based on mica

Metal oxide coated pearl luster pigments based on natural as well as on synthetic mica are available in various particle sizes. The effects obtained differ greatly with the particle size. Using the finest fractions a silky luster is obtained with good hiding power. Medium particle sizes produce a brilliant, already slightly structured luster with a slightly reduced hiding power. The pigments with a coarser particle size are almost transparent and lead to a glitter effect, because the eye, in close-up, can recognize the individual pigment platelets. Pigment selection should be made according to the desired effect, in consideration of the hiding power of the pigments and the resulting influence upon any additional colorants to be used. Pigments with a good hiding power, that means pigments with an intensive light scattering capability, also possess a high lightening power.

Effect pigments based on synthetic mica, which are available on the market since some time, are characterized by a brighter, nearly white mass tone, which can be seen especially in silver white colors.

Effect pigments based on aluminum oxide

This pigment type, which is for sale since some years, plays till now a minor role in the coloration of thermoplastic materials. The so called "crystal effect" which is very appreciated in coating applications cannot be observed in plastics. The reason for this is the difference in the viscosities of lacquers and plastic melts in the application and the time for immobilization (e.g. the orientation of the pigment platelets and their immobilization in the plastic melt in the mold is a very fast process, while in comparison the drying of a lacquer is a much slower process).

In plastics the most important advantage of this pigment type against pigments based on natural and even synthetic mica is the very high degree of whiteness.

In the processing of plastics the "crystal effect" or sparkling effect can be achieved by using pigment types with larger particles (e.g. based on mica or borosilicate flakes). That is one of the reasons that pigments based on aluminum oxide do not play an important role in the pigmentation of plastics.

Effect pigments based on borosilicate flakes

Pearl luster pigments based on borosilicate flakes (glass flakes, glass platelets) are outstandingly suitable for the pigmentation of plastics. One property which is very important for the performance of a borosilicate flake is the thickness of the flake because this has a big influence on the number of pigment particles in a mass unity. The glass platelets for these pigments were produced in very special procedures (see Chapter 2.10). There are possibilities to produce platelets in different thicknesses and thickness distributions. For the use in plastics pigments on thin glass flakes (about 1 µm thick) are particularly suitable, because these pigments were mechanically stable and efficient [11]. Certainly pigments based on thicker glass flakes (about 5 µm) can be utilized, but then the amount of pigment needed is much higher. That means in practice a silver white pigment with an average glass flake thickness of about 1 µm has about a five time higher productiveness than a pigment which was produced with platelets with an average thickness of about 5 µm.

The thin, in part very coarse borosilicate pigments already show with 0.01% in plastics unique frost effects. At this also the extremely white mass tone and the very smooth surface of this pigments takes effect.

The very white mass tone and the very low scattering power of these pigments do nearly not affect the mass tone of colored plastic materials. These pigments have at low concentrations a very low lightening power.

Effect pigments based on SiO_2-flakes

In this group of transparent interference pigments there is a number of very interesting pigments because of there coloristic behavior. Nearly all this pigments show a distinctive color travel, that means, this pigments caused by their physical three layer composition show an angle dependent color characteristic (see Chapter 2.11). This gives this pigment group a specific attractiveness, they show used alone or in combination with transparent colored pigments a tremendous number of color variations. However, here it has to be considered, that the color travel is reduced by the additional use of hiding pigments or the use in translucent plastics.

Because of the very thin substrate as well as the very thin TiO_2- or Fe_2O_3-layer the number of pigment platelets per mass unit is considerably higher than with the other pigments of the layer substrate type. That means, for the application in plastics about half of the concentration usually necessary for mica based pigments is sufficient to achieve very attractive effects.

The following details apply to all above-named effect pigment types, whereas the special properties of the particular pigment have to be considered in addition.

The classic color effect of pearl luster pigments is the silver white luster, known everywhere from the presence of natural pearl. This luster can be shaded with (mostly) transparent colorants. In this way one can introduce a light luster to every color.

Small additions of carbon black to silver white pearl luster pigments can create color effects ranging from silver gray shades and aluminum tones through to antique silver effects. With very coarse pigments even nearly black colors with a glitter or sparkle effect can be created.

The colorfully shining pearl luster pigments, the interference pigments, give to plastics a unique appearance, which cannot be created by other means. The color image changes, which occur with the angle of observation, are called color flop or color travel effect. Thus when combined with colorants, effects can be observed which demonstrate a strong color flop.

Interference pigments combined with carbon black demonstrate the reflected interference color only, along with gray to black shades (out of the gloss angle). It almost appears as if the interference pigments have a surface color themselves.

The mixing of interference pigments leads in nearly always to weak colors and low luster. The mixture of interference pigments with complementary colors leads, depending on the mixture, to gray shades with a certain tint. The only exception of this recommendation is the mixture of two or more very coarse interference pigments in particular pigments based on glass flakes, which create attractive glittering multicolor effects, which can be observed in short distances. Needless to say these mixtures of coarse interference pigments can be combined with other colorants, in particular with carbon black achieving pleasant color effects. When using iron oxide containing interference pigments, the gold and metal colored luster pigments, effects ranging from gold, bronze, copper to red luster effects in almost every shade can be realized. In combination with small amounts of carbon black, even antique metallic colors can be created.

Pearl luster pigments are generally used in thermoplastics in a concentration level of approx. 0.1 to 2 %. In films or in thin layers, considerably higher concentrations are required to achieve an equal effect to that of thicker or solid articles. Naturally, the pigment concentration in the end product is determined by the desired effect. When using a very low concentration of pigment particles (below 0.1 %), particularly with the coarser pearl luster pigments where the individual platelets are recognizable, a similarity to the formation of ice crystals on glass can be found in the final product. This effect is also called the "frost effect", because a "cool" image is produced.

All the above mentioned pigment types have their own optimal application concentration which differs from pigment to pigment. The following remarks will help: Pigments based on SiO_2-platelets and thin glass flakes are very efficient, while pigments based on mica, natural and synthetic one, and on Al_2O_3 need a little higher pigment concentration. The cited concentrations should be understood in the following sense, the lower apply for the more efficient pigment, while the higher ones apply more to the mica based pigments.

4.3.3 Marble effects

Marble effects are a valuable design element in the injection molding process of thermoplastic molded materials. Using pearl luster pigments, the fine crystalline structure of marble is more pronounced, the pigments make the effect more "lively".

A marble effect can be achieved using different methods. The most common procedures are:

- injection molding with so-called marble units,
- injection molding with intermittent injection molding machines,
- plunger type injection machines (very few of these old machines are still in use).

But marble effects can also be produced without modifications of conventional injection molding machines, for example by using plastics with different melt-indices. All of these common methods permit an application with pearl luster pigments.

The last method mentioned above can present some problems; it works only with some materials. An exemplary material combination that produces interesting marble effects on conventional machines, consists of two mixable polystyrene types with different MVR (melt volume rate according to DIN 1133):

MVR 1 = 9.0 cm^3/10 min (200 °C, 50 N) for low viscosity PS

MVR 2 = 1.2 cm^3/10 min (200 °C, 50 N) for high viscosity PS

Pearl luster pigments can be used in either the high or low viscosity components or in both PS types. The pigments must already be worked into the individual plastics (as a compound [12]). If pearl lustrous bands are desired, the pigment concentration in the higher viscosity material (slow flowing) should be considerably higher than in the lower viscosity mixture.

Pearl luster pigment concentrations of 5 to 10 % are quite realistic in the higher melting component. Of course, the percentage of high viscosity material in the mixture must be taken into account when selecting the pigment concentration. Variations of the pigment type and concentration in both compounds and the unlimited selection of the mixture ratios provide a wide variety of marble effects.

4.3.4 Application procedures for pigmenting thermoplastics

Several procedures are available for the pigmenting of thermoplastics (molding materials); these are described in the next chapters. In the industry, the use of a pearl lustrous masterbatch plays a predominant role. A masterbatch (MB) is a pigment preparation, in which the pigment is much more highly concentrated than in the final product and is in a fully dispersed form within a matrix, which is compatible (usually chemically similar) with the application material [12]. Direct coloring is a simple method, but plays a minor role because it does not fulfill all the requirements placed upon the pigmenting quality. Occasionally agglomerates and cords may appear.

As mentioned before, effect pigments are sensitive to mechanical stress, this has to be considered while processing. Investigations have shown, that the major damage of the platelets happens in the steps where pigment powder and unmolded plastic are side by side and agitated (this "ball mill effect" in the production of the premix and in the cold feeding zone of the processing machine) and in the time where the plastic component is still in a very high viscous condition and very tacky. If the pigment is completely wetted and inside the plastic melt, under the conditions of the following plastic processing, only a slight additional breakage is observed.

That means for the processing parameters of the machines:

- keep the mixing time for the premix short as possible and constant and
- in the plastic processing equipment try to reach the melting range of the plastic material and exceed it preferably fast.

This is especially true for high melting plastics like polycarbonate and polyamide.

Direct pigmenting of thermoplastic molding compounds
Pearl luster pigments are easily dispersed. They can be introduced directly into the molding compound before processing and processed together in the machinery. The pigmentation procedure varies depending upon whether the molding material is a powder or pellets.

Direct pigmenting of plastic powders
The direct processing of plastic powders to the final product is usually used in the manufacturing of PVC only. Other molding materials are almost always available as pellets. Simply mixing plastic powders uniformly with pearl luster pigments is sufficient to color them. Tumbling or high-speed mixers are suitable. When using high-speed mixers, the mixing time should be kept short to prevent destroying the pigment platelets. If other additives, colorants or plasticizers are added to the high-speed mixer (for example while compounding PVC) they should be thoroughly mixed in first to the plastic powder and the pearl luster pigments added shortly before the end of mixing time.

Direct pigmenting of the plastic pellets
During processing of plastic pellets directly with the dry pearl luster pigments the components usually tend to separate. A good method, which has proved itself quite well, is to adhere the pigment on the plastic granulate with the use of a compatible fluid substance. Plasticizers or highly refined mineral oils, which contain wetting additives such as emulsifiers or high boiling solvents, are suitable for this. The necessary amount of adherent lies between 0.1 and 0.5 % as a proportion of the plastic.

If additional color pigments are to be used, it is recommended to use pre-dispersed pigment preparations (e.g. as plasticizer-containing paste) and to dilute them in some of the adherent.

In a first step, the plastic pellets were wetted in a mixer (e.g. a tumbling mixer) with the wetting agent. Then the effect pigment can be added and mixed for about 5 min (depending on the mixer) until a homogeneous distribution is achieved. After this short mixing, the pigment "adheres" itself to the granules. A homogeneous pigmentation of up to about 2 % pigment supplement can be achieved in this manner because the surface of the granules is covered, so that no more pigment platelets can stick and will separate. Granules with a small granulate size (< 3 mm) are more advantageous for good pigmentation.

This method is particularly suitable for low viscosity, injection-molding compounds. If the pearl luster pigments still do not distribute uniformly, it is recommended to pass the granulate/pigment mixture through an extruder and then to granulate. This so-called compound, a plastic granulate produced in this way, contains the pigments in the correct concentration and can be then directly processed.

4.3.5 Pigmentation of thermoplastics with pigment preparations

Pigment concentrates, which use a carrier of identical chemical basis as in the final application, such as masterbatches (MB), lead to the best results when coloring with pearl luster pigments in all thermoplastic processing methods and for all thermoplastics. The dispersion of pigments in the MB is generally very good. Some problems occasionally arise in further processing, if the mixing efficiency of the processing machines does not provide a complete distribution within the matrix. Another cause could be a poor viscosity relationship of the masterbatch with respect to the plastic material to be colored (melt viscosity of the MB is too high).

Two methods are available for the masterbatch production, a continuous and a discontinuous method. For industrial production of MB's, a continuous processing with twin-screw extruder is mostly used. The design of the screw should be adapted to the task: disperse the pigment platelets without breaking too much of them through mechanical stress. The difficulties grow with the increase of pigment concentration in the MB. Today, masterbatches with a 30 to 50 % pearl luster pigment concentration are quite common.

The best way to produce high quality MB is the combination of a twin-screw extruder with a side feeder. During this process the effect pigments or suitable effect pigment preparations are inserted to the liquid plastic melt using an additional drive system. Advantages of this procedure are:

- plastic pellets can be processed instead of powder
- very smooth incorporation of the effect pigments, minor mechanical stress for the platelets
- no premix necessary
- simple, flexible mode of operation.

But these advantages are faced with higher investment costs for the equipment.

While processing pearl luster pigments normally creates dust, therefore special pigment preparations were developed for the use in polyolefin, one of the main application fields of pearl luster pigments, with a strongly reduced dusting and at the same time have an excellent influence on the throughput in the MB production.

These preparations in powder form or as granules contain up to 70 % of pigment in low molecular polyethylene or polyethylene wax. The higher bulk density of these products in comparison to the pure pigment allows a considerable higher throughput in a twin-screw extruder. An increase of the throughput depending on the working conditions up to six times was found [13]. In addition these pigment preparations show an improved flowability which prevents hopper plugging.

Even on single screw extruders with these preparations high concentrate pearl MB's can be produced in a very good quality which is frequently impossible with pearl pigments in powder form. For example on a certain single screw extruder with a conventional TiO_2-mica pigment in PE-LLD a MB with a concentration of not more than 10 % was achievable. But when using such a pigment preparation with 70 % of the same pearl pigment a MB with 40 % pigment loading could be made without any problem.

Masterbatch production

To manufacture a masterbatch one preferably begins with the polymer in powder or in fine particulate form. The carrier material should have a high melt volume-flow rate (MVR according to ISO 1133), that means low viscosity (easily flowing), since the addition of pearl luster pigments leads to a lowering of the MVR in proportion to the pigment concentration.

Masterbatch production using PE as an example

PE-LD or PE-LLD powder is generally used for the production of a PE-masterbatch, with a MVR ranging from 5 up to 200 cm^3/10 min (preferably between 20 and 70 cm^3/10 min). The necessary stabilizers for the polyethylene are generally already incorporated in commercially available PE-LD powders, which are made through grinding of PE-pellets.

Occasionally polyethylene wax and/or paraffin oil is added to improve pigment wetting and the flow properties of the MB. Other additives necessary for the final product are frequently incorporated into the masterbatch (UV-absorbers, static inhibitors, additional anti-oxidants, etc.).

Procedures for producing a 30 % PE-masterbatch

a) with a twin-screw extruder

70 parts PE-powder are pre-mixed with 30 parts pearl luster pigment. The mixing time varies with the size of the batch and the intensity of the mixing equipment, roughly 3 to 10 min. Thereafter, the powder mix is as customary introduced into a twin-screw extruder and extruded while at the same time maintaining the temperature profile (approx. 140 °C at the entrance and 160 to 180 °C at the die). Granulation can be done by hot-face cutting or cold pelletizing.

b) with a two roll mill or kneader (preferred for laboratory and small quantities)

70 parts polyethylene type (pellets, grains or powder) are processed with 30 parts pearl luster pigment in a roll mill or a laboratory kneader at a temperature, usually between 150 to 170 °C, until a homogeneous mixture is achieved (approx. 10 min). Pre-mixing of the plastic and the pigment is not obligatory. For a most careful introduction of the pigment, it is advantageous to melt the plastic first and then add the pigment to the melt. Using this procedure significantly lower amount of broken pigment particles is generated compared to the combined addition of pigment and polymer.

After removing it from the machine and cooling it, the mixture can then be reduced to a finer form by milling. The resulting milled material has a broad particle size range and is therefore more difficult to accurately dose than it is with the regular shaped cylindrical or lentiform pellets. During further processing it is important to observe that the MB and the uncolored plastic pellets do not separate in the hopper of the machine. Should this happen it would possibly result in the end product having a non-uniform appearance.

Pigmenting plastic pellets with pearl luster pigment pastes
Pearl luster pigments can also be used in liquid color concentrates (pigment pastes). During the storage of the liquid color, the relative large pigment platelets tend to form hard sediment; this should be taken into consideration. The formation of hard sediment should be avoided. Helpful in this aspect are both the application of a thixotropic additive or maintaining a certain amount of flocculation. Since the pigment concentration is only 40 % at the most, pearl luster pigment pastes are not very suitable for coloring thermoplastics this way. Especially, as too much low molecular material is introduced into the plastic by the fluid phase, which can lead to problems. Unfortunately, with pigment concentrations above 40 %, one cannot achieve a free flowing paste. For this reason the application of pearl luster pigment pastes plays a very minor role.

4.3.6 Special cases of pearl luster pigmentation

Although pearl luster pigments can be used in practically all thermoplastics, there are certain materials where attention to various details must be observed to achieve fault-free results.

Polyolefins
Polyolefins must be stabilized before they are used in the practice. Phenol derivates are frequently used as anti-oxidants for polyolefins. The use of titanium dioxide pigments in polyolefins containing phenol derivative anti-oxidants can lead to an interaction between the anti-oxidants and the titanium dioxide, resulting in the yellowing of the plastic. Depending upon the type of phenol derivative used, yellowing may develop during processing (derivative with a week steric hindrance) or during dark storage of the final products (phenol with a strong steric hindrance) [10].

Normal titanium dioxide pearl luster pigments are sufficiently stabilized for everyday applications, so that neither during processing nor during dark storage does a disturbing yellowing develop. However, it should be kept in mind that the type of the phenol derivative and the quality of the polyolefin has a considerable influence upon this. For highest requirements there are available stabilized pearl luster pigments, which reduce yellowing to a minimum both in the processing and in dark storage [14, 15].

Polyvinyl chloride

Polyvinyl chloride, which differs slightly from most molding plastics during compounding, can be colored with pearl luster pigments in both plasticized (PVC-P, old terminology: soft-PVC) and un-plasticized form (PVC-U, old terminology: hard PVC). The luster effects are the more brilliant, the more transparent the PVC compound is. Therefore, while stabilizing the PVC, it is important to maintain the transparency of the stabilized PVC-compound. Compounding can damage the pigment platelets due to too aggressive mechanical forces, which in turn will bring about a considerable loss in luster. For this reason pearl luster pigments should be introduced shortly before the end of the compounding process, but still in the hot mixer. All metal oxide coated effect pigments can be used in PVC's. In some PVC compounds however, the presence of iron ions may lead to stability problems (light, weather and temperature resistance). Before using iron oxide containing pearl luster pigments, pre-testing should be carried out for suitability.

Drying problems

Plastics which must be dried before processing (for example PMMA, PC, PA etc.) require special preparation steps. It is a fact that metal oxide containing effect pigments contain small traces of absorbed water as delivered (less than 0.5 %), which can increase greatly if improperly stored. Therefore for moisture sensitive plastics both, plastic pellets as well as the effect pigment or the masterbatch should be predried to an acceptable moisture level before processing. Instructions for drying can be found in the manufacturer's information for the corresponding plastic, which is also valid as minimum drying requirements for the pigment. Even during the production of masterbatches for moisture sensitive plastics, it is recommended that both the final ingredients as well as finished masterbatch are carefully dried. Certainly the suction vent of a twin-screw extruder should always be used to demoisturize the melt under vacuum.

Special case polycarbonate

For high quality applications of pearl luster pigmented polycarbonate the influence of OH-groups especially on the surface of TiO_2 coated effect pigments which still exist after the recommended drying procedure has to be considered. These OH-groups can degrade the polycarbonate under the conditions of MB production and further processing in such a way that the properties of the material may become crucial. The degradation reactions reduce the molecular mass and the color turns more yellowish. For this reason the mechanical properties (e.g. impact strength) degrade too. In such applications the addition of appropriate additives is necessary. These additives have to react with the OH-groups on the pigment surface and as a consequence the degradation of the polycarbonate is reduced. Practical experience showed that H-siloxanes are effective even in low concentrations, depending on the type of H-siloxane, 1.5 to 2 % based on effect pigment is sufficient. In any case the recommendations of the producer have to be considered while processing. This reaction generates hydrogen which has to be removed safe. The H-siloxane is processed together with the pigment while the production of a MB or a compound.

Plastics for outdoor applications

Special stabilized pearl luster pigments have been especially designed for the coloration of plastics that are designed for outdoor applications, for example PMMA, PVC, PE, etc. In titanium dioxide mica pigments, the stabilization is there to reduce the photo catalytic properties of the titanium dioxide surface and simultaneously to improve the interaction between pigment and the plastic matrix.

In the case of iron oxide mica pigments, only the interaction with the plastic matrix is improved, since iron oxide is not photoactive.

In some (rare) cases, good weather resistant results were achieved with standard pigments, but these results can be greatly improved with additional stabilization of the pearl luster pigments.

In these applications, the weather resistance of a pigmented plastic is a property of the total system – plastic, plastic stabilizers and pigments together. Not only the individual components should be examined for their weathering properties, but also the entire system should be tested in the field. Usually the weathering results observed for such a system cannot easily be transferred to another system. Such conclusions are especially difficult for pearl luster pigmented materials.

4.3.7 Processing pearl luster molding materials and recycling

Because pearl lustrous pigments, as described, can be used in virtually all thermoplastic molding materials and in many cases they already are, the entire spectrum of thermoplastic processing is available for applications with pearl lustrous plastic materials [16]. The most important methods for plastic processing are extrusion [17–20] and injection molding [21, 22]. Both methods are also suitable for pearl luster colored molding materials.

Special cases are the plastisols or organisols. Especially PVC plastisols are in use to produce soft hollowware through centrifugal casting or artificial leather by coating woven fabrics (balls, toys for babies, artificial leather). In this processes pearl luster pigments can be applied with good results. But it is important that at a particular time while or after gelification a shear flow in the material is created to orientate the pigment platelets properly. If this does not happen, the orientation of the effect pigments is moderate and the resulting effect is low.

The incorporation of the effect pigments preferably done by wetting out the pigments with plasticizer in the relation of 1 part pigment to 2 parts of plasticizer before adding to the plastisol or organosol. In this way the introduction of air bubbles is very low.

Another special case in the processing is the rotational casting [23]. In this process mostly high viscous polymer powders, especially polyolefin powders, were melted in heated molds which are rotating in three directions into a uniform layer. After cooling down in general one have a thick walled and if necessary quite big hollow pieces. Unfortunately in this process the flow of the melt is so marginal that the orientation of the effect pigment platelets in the melt is insufficient. The effects achievable do not match in no way the effects known from the other thermoplastic processes. For this application process preferably large sized, very brilliant, pigments are suitable (e.g. glass flake pigments). The effect pigments need not be incorporated in the plastic powder, a dry blend of pigment and plastic powder is suitable.

Extrusion

In many processing methods extrusion is but one step towards melting, homogenizing, and pressuring ready for ejecting the material, which is then further processed into granulates for the production of blown hollow bodies, blown foils, flat foils, sheets, tubes, hoses, profiles, etc. Most important for pearl luster materials at this time is the extrusion blow molding process.

In almost all extrusion processes, the use of masterbatches or compounds is unavoidable in order to produce a fault-free coloring of the final product.

The possibilities with this method have expanded greatly in recent years through the development of the co-extrusion technique. In this process, a multi-layered wall construction can be achieved. A particular function can be delegated to each layer. The outer layer, for example, serves the design, while the middle layer offers the possibility to use in-plant recycled plastic materials. The requirements of the inner layer depend upon the constituents of the contents.

These newer methods offer interesting perspectives for pearl luster pigment design applications. They are found almost exclusively in the outer layer, the decorative layer. During co-extrusion, the colored pearl lustrous pigment layer is considerably thinner than in normal extrusion. This must be noted while calculating the pigment concentration. Depending upon the layer thickness, the pearl pigment concentration can be in the order of 10 %.

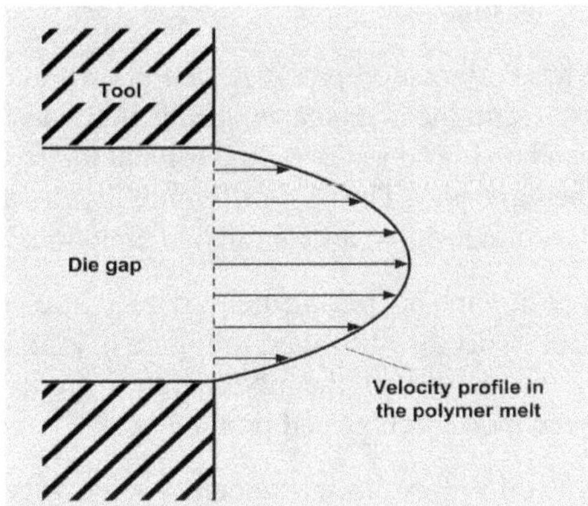

Figure 4.4: Schematic illustration of the orientation processes in the die gap during extrusion

Generally, processing with pearl luster molding materials is done using normal machine settings. But, as with all materials, the optimal processing parameters should be established and maintained.

In the extrusion processing the pigment orientation occurs as the melt passes through the orifice of the die. This process is illustrated in Figure 4.4.

It is easy to understand how surface irregularities or dirt particles that hang-on or hang-up in the die gap due to a poor tool design can lead to a non-uniform flowing and thus to disturbances in the final pigment orientation. Careful design and machining of the tools, as well as treatment and maintenance are of absolute importance.

Injection molding

A further important processing method for pearl lustrous pigmented plastics is the injection molding process [21,22].

Pearl luster pigments can be used directly as a powder in a number of plastics as described in Chapter 4.3.4. Of course also masterbatches are suitable if an excellent coloring quality is crucial. Fluid coloring is used less frequently with pearl lustrous injection molded parts. Compounds can also be used, particularly when a very good and homogeneous pigmentation quality is required.

Plastics colored with pearl luster pigments are best processed at temperatures lying in the upper end of the manufacturer`s recommended temperature processing range whereby the luster effects will be better and more uniform.

The other machine parameters should be adjusted to the recommended optimal values for the plastic material being processed. The pigment platelet orientation in injection molding takes place in the flowing processes during molten phase of the plastic.

These orientation processes are known to occur both at the molecular level [24] as well as in the fiber shaped filling materials [25,26]. The pearl luster pigment platelets can be viewed as “two dimensional fibers”. The orientation takes place always in the same way. The only difference is that the orientation of the pigment platelets is two-dimensional and not in one direction, as is the case for fibers.

The moving hot plastic melt front has a velocity profile, which is created, within the tool cavity, as illustrated in Figure 4.5 (a). Somewhat behind this (in the direction of the melt flow), the modified flow profile (b) can be found, still further back an even more pronounced profile (c).

The differences in the velocity profile arise because it is cooler at the walls of the mold, which in turn increases the melt viscosity in these regions. This increases the viscosity until the melt is no longer mobile (solid). Similar to other described

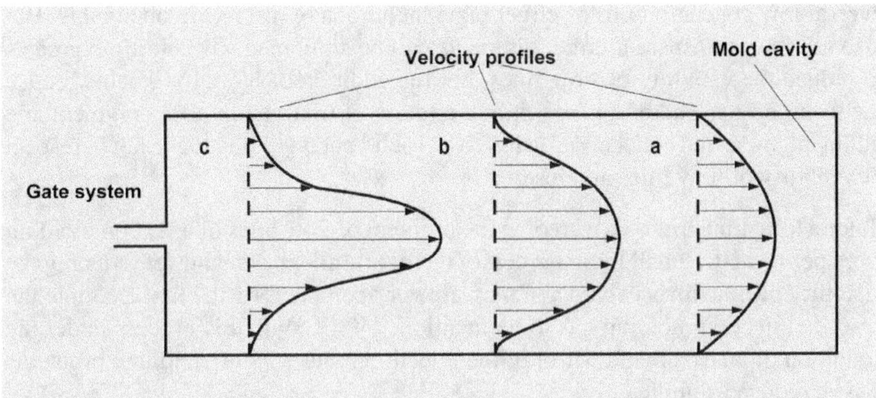

Figure 4.5: Schematic illustration of the orientation processes in an injection mold

orientation processes, the velocity drop in the melt leads to an orientation of the platelets. Near to the surface, the platelets align themselves nearly parallel to the surface. Theoretically, the platelets lie perpendicular to the flow direction in the middle of the mass flow and on the flow front. Occasionally there are additional currents like elongation flow that have to be considered.

Combination of all these forces in an undisturbed flow pattern will produce a good orientation of the pigment platelets.

Depending upon the design and shape of the mold as well as the injection position and type, same noticeable flow lines and/or weld lines can develop. The platelet structure of the pearl luster pigments enhances the visible effect. However, these flow lines also quite common in normal injection molded plastics, sometimes more or less noticeable but are almost always detectable.

With an optimal design of the mold shape, the tooling of the mold, and the gate, flow paths and the correct mounting of the mold parts the flow and weld lines can be reduced to an absolute minimum. Even the type of pearl luster pigment selected can have an influence upon the obviousness of the flow lines. Owing to the transparency of the large particle pigments their influence upon visible disturbances is considerably less than with the pearl luster pigments of a smaller particle size. But in addition the total color formulation including conventional pigments play an important role. The proper recipe reduces the visibility of flow lines, but the use of hiding pigments reduces the brilliance. Here the combination of very large particle size, high brilliant pigments based on glass flakes and hiding pigments have to be mentioned. As hiding pigments titanium dioxide pigments and other inorganic pigments can be used. As mentioned before glass flake pigments with very thin platelets (less than about 1 µm) and a high brilliance are very efficient and even at low concentrations of effect pigment attractive results are obtainable. But also all other transparent effect pigments can be combined with hiding pigments to reduce the visibility of flow lines, but the addition of hiding pigments reduce the luster effect. Therefore the optimal relation of transparent effect pigment and hiding pigment and concentration has to be found out regarding the result in respect of visibility of flow lines and luster.

Since a long time trials to overcome the problem of flow lines in injection molding were performed. But all these processes did not fulfill the demands of reducing the visibility of flow lines even when they are very complex. As the first example the cascade injection molding has to be mentioned [21, 22]. With this process under the condition of an excellent control of the injection times a lot of weld line problems can be reduced significantly.

Another example is the so called pulse-push-process (Figure 4.6) [27]. The basis of this process is the idea that as long as the plastic material in the mold is still liquid through pumping the melt back and forth a better mixing of the two melt flows should be possible. This is achieved by a control element (e.g. a rotary valve) which divides the melt coming from the nozzle of the injection molding machine into two or more melt flows which are injected into the mold through different gates. In some simple molds this concept could be proved but at more complicated molds and parts results are not yet on hand.

As a third example the Scorim-process has to be mentioned [28]. According to this process the melt in the mold is moved by means of a hydraulic plunger while injection and after-pressure of the injection molding process. This process needs two gates. The motion of the plunger causes a back and forth flow of the plastic core and as a function of this an orientation of fibers, platelets and molecules in the material. Although this process is relatively old it is used only rarely.

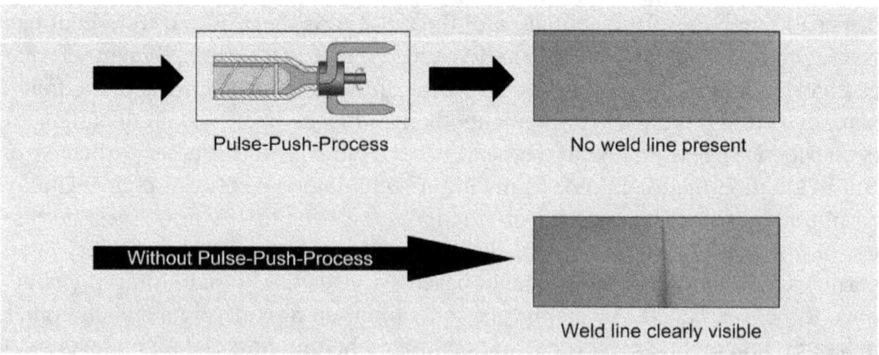

Figure 4.6: Schematic illustration of the pulse-push-process

Recycling of effect pigment containing plastic

All the before mentioned effect pigments generally do not influence because of their chemical, physical and toxicological properties neither the chemical nor the material recycling or the energy recovery of plastics waste.

But in some very special applications some problems may arise if effect pigment containing plastic material (e.g. polyethylene terephtalate, PET) should be transformed into fine fibers. In this process the relative large pigment particles may plug the spinning nozzle.

4.4 Pearl luster pigments for laser marking

The marking of production articles using laser marking technology is expanding rapidly into many new applications and application areas (see Chapter 2.8.7.3). Due to the high quality of the marking or images possible today, the technology has become interesting not only for product identification, expiration dates, warnings, etc., but also for decorative and design elements of the end product.

Next to the sharp lettering of the print quality, the laser technology offers further advantages. The writing speed is extremely high and the marking is immediately permanent and indelible. It is an ink free marking process. Thin film contaminants, such as moisture on the surface of the object do not influence the marking. There is no mechanical wear and tear to the equipment or the object (target) because the marking is successfully done in a non-contact manner. Thus maintenance and operating costs can be considered as relatively negligible. Since the marking is performed directly on or within the object to be marked, it is permanent, indelible, and practically non-alterable (good for product security).

In order for the marking to be highly legible, the laser type employed should be able to initiate a marking reaction within the plastic, which is both adequately contrasting and irreversible. These pre-requirements are only sufficient in a few types of plastics. The majority of plastics are able to accept a weak type of marking or there is no marking to be seen, in the sense that the contrast is negligible or it appears as an etching without a color change (no contrast).

In many cases by the addition of a small amount of pearl luster pigments, based upon oxide coated mica, the desired contrasting reaction can be achieved. Sharp markings and images within the surface area of the material are the result, as illustrated in Figure 4.7.

Figure 4.7: Laser-marked laptop-keyboard (Nd:YAG-laser)

4.4.1 Functionality of laser-sensitive pearl luster pigments

The functional mechanism when using mica-based pearl luster pigments can be described as follows: The platelet shaped pigments absorb the incidental laser light and become themselves a temperature sink, which, in turn, means that the temperature of the pigment particles increases rapidly. In many materials this will lead to a carbonization of the polymer material surrounding the pigment; resulting in the plastic turning a darker color specifically at the laser-targeted point.

Other plastics tend to exude reaction gases at the elevated temperatures created by the laser, whereby, they exhibit a light to white colored marking due to the entrapment of extremely fine gas bubbles.

Carbonizing and the exuding of fine gas bubbles can appeal simultaneously, thus producing light gray markings. The relationship between carbonization and fine gas bubble build up is not only dependent upon the plastic type, but also upon the laser energy intensity.

The tendency of a plastic material to carbonize or to foam up, that means to give dark or light markings, is more or less a property of the plastic type which can be influenced only a little bit by additives. These reactions only take place with the pigments that are near to the surface of the plastic object; hence, there is virtually no influence upon the properties or the stability of the plastic material.

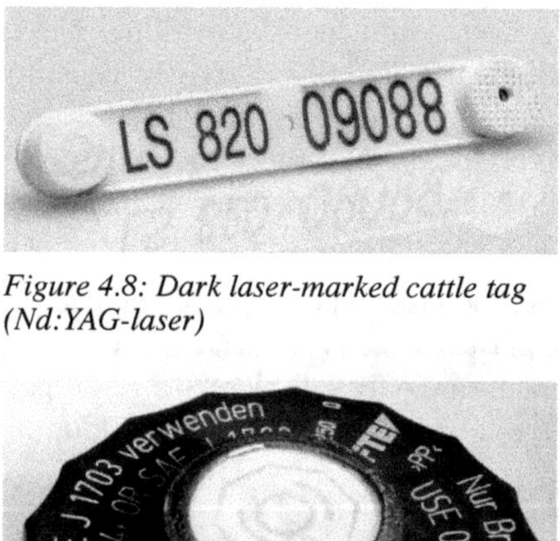

Figure 4.8: Dark laser-marked cattle tag (Nd:YAG-laser)

Figure 4.9: Bright laser-marked closure of a brake fluid tank (Nd:YAG-laser)

The selection of the laser-sensitive additive is normally only depending on the type of laser used, because the effectiveness of the additive is depending on the wavelength of the laser. Some examples for laser marking with an Nd:YAG-laser are shown in the Figures 4.7 to 4.9.

4.4.2 Laser marking methods

Currently three types of lasers play an important role in the marking of plastics:

- CO_2-laser (wave length mostly 10.6 µm) pulsed and as a dot matrix
- Nd:YAG-laser (wave length 1064 nm or 532 nm with double frequency)
- Excimer-laser (wave length depending upon the gas type, between 193 and 351 nm).

When writing with these lasers three methods are in use.

With low frequency pulsed CO_2-lasers and excimer-lasers, the marking is achieved by a masking method. With high frequency pulsed or continuous CO_2-lasers and the Nd:YAG-lasers, a deflected ray method is used by means of the use of mirrors. The dot-matrix process can be regarded as a variation of the deflection ray method.

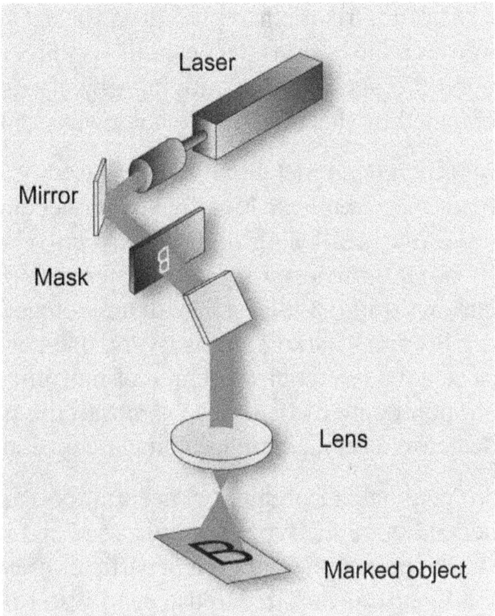

Figure 4.10: Scheme for laser marking using the masking method [29]

The masking method (Figure 4.10) requires an extremely short pulse of light energy to be directed at the object to be marked. In this case the stencil lettered metal mask is situated in the path of the laser beam. Thereafter, the emitted (masked) laser beam passes through a lens, which is focused upon the object to be marked. By varying the projection strength, the energy density in the laser beam (in J/cm) can be further increased.

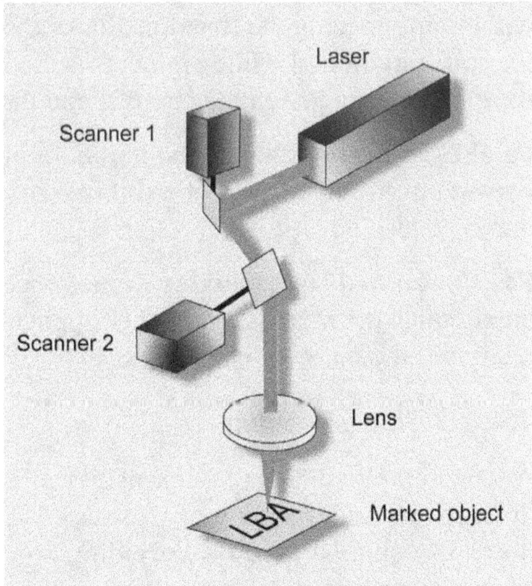

Figure 4.11: Scheme for laser marking using the beam deflection method [29]

In the beam deflection (galvanometer) method (Figure 4.11), the laser beam is guided to the object to be marked by the use of two mirrors (galvanometer type) and a lens system. A computer with specific software programs is used for the marking required and controlling the movement of the mirrors. The marking occurs during the deflection of the laser beam in both the x- and y-directions. The beam deflection method is well known for its flexibility, and that it is able to transmit to the object a high density of information. Nevertheless, the marking speed is not as fast as by the masking method.

The dot-matrix method presents a special variation of the beam deflecting method, in which the beam is deflected in one direction only. The second direction depends upon the speed of the moving object passing by the writing head. The results obtained by this method are not always optimal.

In principle all metal oxide mica based pearl luster pigments are suitable for laser marking. Even very low pigment concentrations are sufficient to enable a plastic to be marked. Using high-density polyethylene (PE-HD), at a concentration of 0.1 % TiO_2-mica pearl luster pigment can produce a contrasting and easily legible, dark writing (CO_2-laser). With an increase in pigment concentration, so the writing becomes darker, alternatively, one can adjust the strength of the laser energy to achieve the required degree of uniform darkening (specific contrast). The penetration depth of the laser beam into the plastic and the depth of the writing will decrease with increase in pigment concentration.

An economical pigment concentration for optimal contrasting laser writing will depend upon the type of plastic used and its thickness. Likewise, there must be a sufficient quantity of pigment particles present in the writing area to achieve a good marking. It is possible to use pearl luster pigments (alone or with non-transparent color pigments) in the plastic articles for their laser marking properties without introducing a pearl lustrous effect (pearl color) to the plastic, because highly scattering pigments, such as TiO_2-white pigment, completely hide the luster effect without influencing the functionality of the "laser-active" pearl luster pigments. If the concentration of white pigments is too high, however, it could possibly lighten the marking and lessen the contrast and the legibility.

Parallel to pearl luster pigments, there are also nowadays pigments of similar construction, which do not exhibit pearl luster properties but give good laser marking results.

Pearl luster and laser marking do not necessarily go together. There is still more empirical work necessary in this application field to enable better predictions regarding the marking of plastics.

4.4.3 Laser welding

Another employment of laser sensitive effect pigments is the fast growing application area of laser welding of plastics [30]. The transmission laser beam welding is the most common used laser welding process (Figure 4.12). For this process it is necessary to have a laser transparent and a laser absorbing part in direct contact which are welded by irradiation of a laser beam through the transparent part.

For laser welding diode laser with wavelength of 808 or 940 nm, Nd:YAG-laser with 1064 nm and CO_2-laser with 10.6 μm are in use. The selection of the laser sensitive effect pigments is influenced predominantly by the laser type intended.

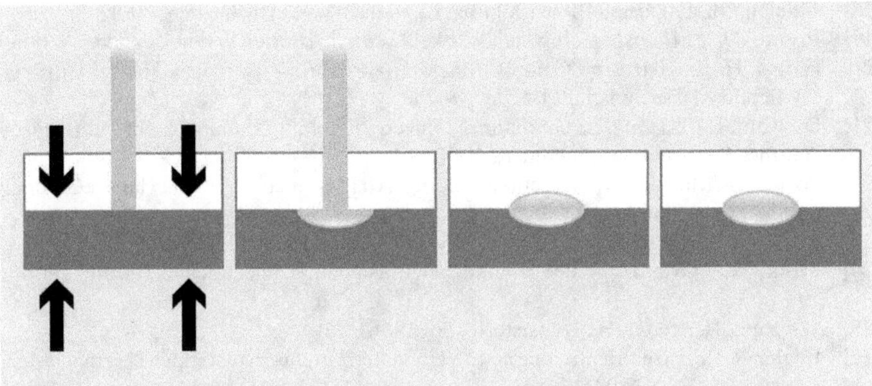

Figure 4.12: Scheme of the laser welding process.

In the laser absorbing part carbon black can be used for each laser, but in this case a loss, it is always black and this make the color design difficult. Therefore the pigments described for laser marking are a good choice for bright and colored parts. The concentration of these pigments is depending on the wall thickness of the join partners in general between 0.2 and 1%.

References

[1] Trojanowski, G., Brandt, L., US 2,962,767

[2] Stawowy, J., in Kunststoff-Handbuch, Vol. 10, C. Hanser Verlag München, 1988

[3] Stahl, D., Strolenberg, O. K., Damen, L. W. J., DE 16 94 516

[4] Eulgem, A., Woinke, R., DE 11 43 024

[5] Gerson, E., DE 14 46 830

[6] Gerson, E., Etlenyi, T., Bruins, P. F., US 3,253,070

[7] Gerson, E., Bruins, P. F., DE 10 83 537

[8] Bäumer, W., "Perlglanz und Perlglanzpigmente", Farbe + Lack, 79 (1973) 747

[9] Kieser, M., Edler, G., unpublished results

[10] Meckenstock, K. U., "Einfärben von Kunststoffen mit Titandioxid-Pigmenten", Expert Verlag, Sindelfingen, 1984

[11] Melson, S., Entenmann, M., Jekel, M., Mathias, M., WO 2006/018196

[12] DIN 55943:2001

[13] Hilder, R., Kieser, M., US 6,451,102

[14] Dietz, J., Kieser, M., Iwasa, K., DE 40 41 663

[15] Franz, K.-D., Dietz, J., Kieser, M., DE 41 21 352

[16] Osswald, T. A., Baur, E., Brinkmann, S., Oberbach, K., Schmachtenberg, E., International Plastics Handbook: The Resource for Engineers, Hanser Gardner Publications, Cincinnati, 2006

[17] Chung, C. I., Extrusion of Polymers: Theory and Practice, Hanser Gardner Publications, Cincinnati, 2000

[18] Thielen, M., Extrusion Blow Molding, C. Hanser Verlag, München, 2001
[19] Rosato, D. V., Blow Molding Handbook, 2nd ed., C. Hanser Verlag, München, 2003
[20] Harris, H. E., Extrusion Control: Machine – Process – Products, Hanser Gardner Publications, Cincinnati, 2004
[21] Osswald, T. A., Turng, L., Gramann, P., Injection Molding Handbook, 2nd ed., Hanser Gardner Publications, Cincinnati, 2007
[22] Pötsch, G., Michaeli, W., Injection Molding: An Introduction, 2nd ed., Hanser Gardner Publications, Cincinnati, 2007
[23] Beall, G. L., Rotational Molding, C. Hanser Verlag, München, 1998
[24] Mink, W., "Grundzüge der Hohlkörper-Blastechnik", Zechner & Hüthig Verlag, 1971
[25] Wiegand, H., Vetter, H., Kunststoffe, 56 (1966), 761
[26] Hegler, R. P., "Faserorientierung beim Verarbeiten kurzfaserverstärkter Thermoplaste", Kunststoffe, 74 (1984) 271
[27] Keim, H.-J., DE 100 52 841
[28] Allan, P. S., Bevis, M. J., US 5,160,466
[29] Corporate publication, Fa. Alltec, Lübeck, 1992
[30] Renner, T., Sieffert, M., "Laserschweißen: Lasergerechte Konstruktion beim Kunststoffschweißen", Kunststoffe, 2 (2004) 30

5 Special effect pigments in printing inks

Joachim Weitzel

Special effect pigments, due to their specific properties, are irreplaceable in the printing industry today. They are indispensable for the shaping of effects and for the presentation of special designs. Because of the great importance of pearl luster pigments in the range of effect pigments for printing applications, this chapter refers in the following especially to this group. Basically the given information is also applicable for all special effect pigments.

5.1 Importance of pearl luster pigments in the printing industry

The demand for pearl luster pigments in the manufacture of packaging, textile materials, wallpapers and furniture facings as well as for high quality papers and carton boards is steadily increasing. One of many advantages of these pigments is that metal-like effects can be generated without using metals.

5.1.1 Design effects

Pearl luster pigments, being highly transparent, can be used, either alone or in combination with all the commonly used transparent color pigments and coloring agents, to create three-dimensional satin effects ranging from a pearlescent sheen right up to glittery sparkle effects [1].

In some cases they are used to conjure up convincingly deceptive natural effects, including wood-grain effects on furniture facings. Depending on the particle size of the pigment as well as its coloristic properties, it is possible to create wood-grain structures that cannot be produced by any other means. Ceramic effects, such as mineral appearance, stone grain structures too, can be incorporated into synthetic floor and wall coverings.

Since interference pearl luster pigments are transparent, and being often used in combination with other coloring agents, or pigments, there is another possibility of creating effects, by incorporating the color of the background, or pre-print. The color of the image will change with the viewing angle ("iridescent" effect), and this opens up a vast range of effect creation possibilities.

In the following some principles are explained. Figure 5.1 (page 172) shows schematically what happens when a red interference pigment is printed on a white background. The light transmitted through the pigment is complementary to the red interference

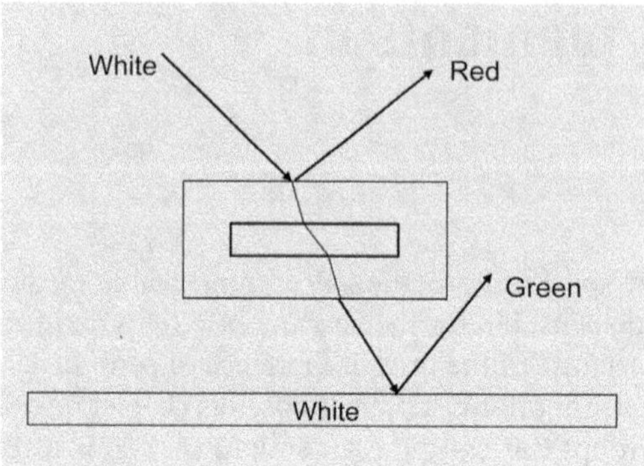

Figure 5.1: Color appearance of a red interference pigment printed on a white background

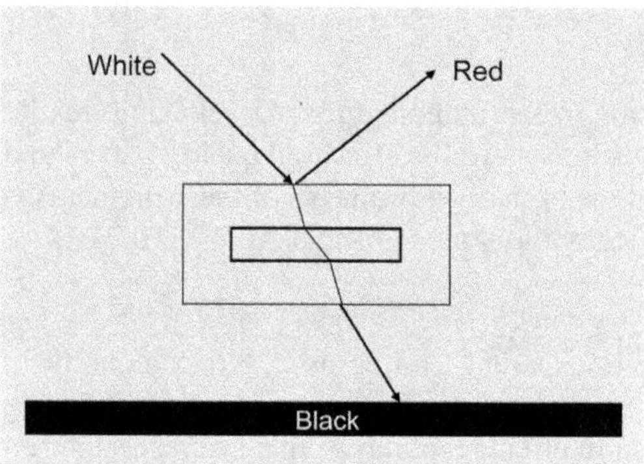

Figure 5.2: Color appearance of a red interference pigment printed on a black background

color and is diffusely reflected by the light background. The print looks brilliant red at the specular angle and green from other angles of observation.

Figure 5.2 shows schematically the effect of printing the same red interference pigment onto a black background. The transmitted light is absorbed and the print will be observed as a lustrous brilliant red.

Many combinations of colors can be realized in this way by printing the individual interference pigments onto a differently colored background or, alternatively, by mixing them together with transparent colors or transparent pigments. The color effect achieved will depend essentially on which one of the two options (background effect or color combination effect) was opted for. The "Iriodin" Color Compendium lists several thousand color and effect variations [2,3].

5.1.2 Environmental protection

Pearl luster pigments have an additional and very important feature, they are environment-friendly, and exceed by far the environmental standards set for the printing industry, as well as meeting the BGA and FDA purity requirements with regard to limits of heavy metals and aromatic amines. They are non-toxic, chemically inert and approved for food packaging applications. These properties are very much consistent with the strict demands imposed on products of the printing industry. The only environmental considerations that need addressing are:

- disposal of residual inks,
- cleaning of printing forms and inking systems,
- recycling of printed products and
- proper disposal of general printing waste.

Pearl luster pigments are completely neutral in all of these processes and, being mineral-based, they do not harm the environment.

5.2 Application areas

5.2.1 Printing onto textile fabrics

For many years now this has been a major market for pearl luster pigments. Textile printing inks containing pearl luster pigments can be used for superb color and pattern effects on substrates ranging from small ornamental objects to a wide variety of textile materials. They are suitable, in particular, for printing onto sports clothing, blouses, dresses, and home-textiles such as bed linen, tablecloths, curtains, wall-coverings, soft furnishings and lampshades.

Pearl luster pigments enhance the overall effect through their vibrant, highly lustrous and glittery appearance. Their water- and alcohol-resistance makes them a popular choice for many applications. Their use for printing onto textiles is particularly apt on account of their excellent resistance to thermo-fixation, solvents, washing, perspiration and chemical cleaning agents. There is no negative influence to the application, when pearl luster pigments are used, due to their high chemical resistance.

Large textile printing houses tend to use pre-formulated (i.e. ready-to-print) pearl luster pigment pastes from big-name manufacturers, supplying large quantities of product at a time.

The principal printing processes used are: rotary screen printing and flatbed screen printing for roll-to-roll printing, and carousel flatbed screen printing for ready-to-wear items such as T-shirts.

In the latter case, solvent-containing, UV drying or 2K screen printing inks are used, for the reason that these inks require much shorter intermediate drying times than water-based dispersions, plus the fact that the carousel printing process offers multicolor pearlescent printing as an option.

Pearlescent pigments can be used in almost all ink systems; however, it is best to use highly transparent systems only. Between 0.5 and 15 wt-% pearl luster pigment should be used, to attain results ranging from rich satin effects to thinner, glittery effects.

Pearl luster pigments can be combined with the usual textile printing auxiliaries. In other words, small amounts (up to 20 wt-% relating to the pearlescent pigment) of any transparent color pigment may be added. The added color components will effect or influence the color shade. Soft, lustrous, silver-, gold-, bronze- and interference-effect results can be obtained. A considerably enhanced effect can be achieved by calendaring or mangling the fabric, which is especially applicable to textile prints and which can be used to produce a high-quality chintz effect.

Pearl luster pigments excel through their brightness in textile pigment prints, through their fastness and their outstanding printing properties. They are also highly suitable for alkali crêpe technique applications and for use in the "burn-out" process where, again, their chemical resistance to the acidic reactions occurring in this process make them an ideal choice of coloring agent for use in combination with reactive pigments.

Their chemical inertness and resistance to acids and alkalis make them ideal candidates for these processes, which involve acidic/thermic reactions to develop colors, or in which basic reactions cause the textile filaments to swell. Pearl luster pigments can be used as an effect within the stop outlines for the application on burn-out materials. This opens up another field of creativity within this very special application.

5.2.2 Wallpapers and furniture facings

Another popular and commercially important application for pearl luster pigments is in the high-quality treatment of papers and other substrates for use in the wallpaper and furniture facing industries. Generally speaking, the same principles apply as for textile pigment printing, with suitable substrates ranging from miniature ornaments and pieces right up to super-calendared surfaces that are further processed by embossing or lamination.

In furniture facings the aim is more to add subtle support to natural structures such as wood-grain. The pigments of choice in this type of application would therefore be silver-white pearl luster pigments which, when used with transparent color pigments, offer almost limitless design possibilities.

The preferred printing process is rotogravure, using water-based dispersion ink systems. However, PVC-based inks are still used, e.g. in the wallpaper industry for creating high-blow and low-blow effects; these are processes in which heat is used to blow up structures pre-printed onto the substrate.

Special characteristics, the pigments must have are good light-fastness and chemical inertness, so as not to interfere with processes such as hot pressing with melamine resins in the manufacturing of furniture facings. The pigments have outstandingly consistent coloristic qualities, which enable users to maintain a high standard of reproducibility.

5.2.3 Fine papers and carton board

This is one application for pearl luster pigments in which the superior properties of these pigments are very apparent [4].

Within the printing of folding carton board for cosmetics, aim is taken on the creation of very strong pearl effects to emphasize on the noble character of the cosmetic product. To achieve this effect, the print must be heavy, with thicknesses that exceed those normally applied in gravure printing. Numerous printing techniques are used, ranging from knife-over-roll coating to screen-printing, and from airbrushing to bronzing. Further coating, overprinting, embossing and hot sealing are additional measures that may be applied to add weight to the nobility of the product.

No matter what process is used, the pigments are easy to incorporate into ink systems, as the layer thicknesses in which they are used, and their anisotropy, allow an optimum lay down over the surface of the substrate, to create maximum effect and luster. There is a discernible trend among end-product manufacturers, printers and design studios towards the use of especially fine papers and fine carton board (pearl pigment coated papers/cartons).

5.2.4 Packaging

The term packaging is fairly broad and embraces materials, such as papers, carton board, plastic film, wood and metal. Pearl luster pigments can be used to print onto all of these materials, subject only to the limitations of the printing process.

The most important materials currently used are printable films and papers:

- film in the form of tubes for sweets and biscuits, and as self-adhesive or shrink labels,
- wrapping paper of 50 to 80 g/m^2, or laminar material for carton board and, increasingly, as a base material for wet-stick labels.

The major processes used for printing on film are rotogravure and flexographic printing.

5.3 Applicable printing methods

5.3.1 Pearl luster pigments in gravure printing

In gravure printing, the print results are formed by the transfer of ink by small cells, etched, engraved or laser engraved into the printing form. Non-printing areas are flat and smooth. The form is either a cylinder or a plate made of steel, copper, chromium or made of a flexible photopolymer.

Full tone areas such as text and solid colors have larger cells, while halftone areas are smaller in size and less deep. Flooding or dipping so as to fill all of the cells, inks the printing form and excess ink is removed from the surface of the form by a flexible blade. This regulates the thickness of the print and assures optimal transfer of the ink. The substrate is fed between the inked printing cylinder and the impression cylinder. The printing area of the form transfers ink to the substrate through pressure and capillary action.

Gravure printing is suitable for high-quality prints and is capable of sustaining high printing speeds for high-volume jobs. Therefore, the technique is used for printing long runs of packaging materials, magazines, catalogues and even some newspapers. As dot transfer requires intimate contact between the cells and the substrate, the best effects are achieved by selecting substrates with even, very smooth surfaces, including fine, highly satined or super-calendared papers, plastic films and sheeting.

A characteristic feature of gravure prints, generated with acid etched, or electro-mechanically engraved printing cylinders, is the jagged edges of lines, text and full tone images, which are visible under a magnifying lens. Printed products, generated with directly laser engraved printing cylinders do not show these characteristics.

Gravure printing presses usually have banks of cylinders and overhead drying units. There are generally up to eight cylinders, occasionally more.

Printing parameters

Pearl luster pigments are generally well suitable for low viscosity solvent or water-borne gravure inks due to their physical and chemical properties.

To achieve the optimum pearlescent effects only highly transparent binder systems should be used. To enhance the incorporation and transfer of pearl luster pigments, an ink vehicle with slightly higher solids is beneficial.

By tinting the formulation with highly transparent pre-dispersed organic or inorganic colorants, a variety of brilliant colors can be created. However, since the pearlescent effect may be somewhat compromised by tinting, a good recommendation would be, using only up to 3 wt-% of the conventional pigments in the ready-to-print ink.

Satin effects can be achieved by using a particle size range of 5 to 25 μm with a concentration of up to 30 wt-% pigment in the ready-to-print formulation.

For enhanced brilliance, particle size ranges of 10 to 60 μm and 10 to 125 μm are recommended using slightly lower concentration levels of up to 25 wt-% in the ready-to-print formulation.

The surface structure of pearlescent pigments contains extremely fine capillaries. Therefore, pre-wetting helps with the incorporation and the homogeneous dispersion, while improving the anti-settling and adhesion properties.

The dry pigment should be wetted out with components of the ink vehicle, e.g. water, alcohol or any other thinner. The resulting pigment paste should soak for a few hours or overnight, and may then be added to the ink vehicle. This procedure also reduces dusting during the final incorporation of the pigment into the ink system.

A new development, in form of pellets, or granulated, pre-wetted pearl pigments is also commercially available [5]. This product range is described in the following within the section of offset varnishing (see Chapter 5.3.6).

For roll-to-roll printing on paper or films, a printing viscosity of 15 to 25 s (DIN-Cup 4 mm) or 20 to 30 s (Zahn-Cup # 2) is advantageous, depending on the printing speed and ink system. For sheet-fed gravure printing as well as wallpaper and decorative paper printing, a slightly higher viscosity of up to 40 s is recommended.

The optimal screen and cell configuration of the cylinder is essential for achieving luster, brilliance and color intensity of the finished print. Pearl luster pigments with a particle size distribution of 5 to 25 µm are the most suitable products for the commonly used 60 to 70 lines/cm cylinders. For enhanced brilliance of a full-tone print, pearl luster pigments with a particle size range of 10 to 60 µm should be used. This particle size range, however, requires a larger screen, e.g. 40 to 60 lines/cm.

When using solvent-based inks with pearl luster pigments it is recommended to use etched cylinders, lasered or engraved cylinders with an opening angle of 120° and a cell depth of about 50 µm. The cell-to-cell wall ratio should not be less than 5:1.

When using waterborne inks with pearl luster pigments the choice should be on engraved cylinders with an opening angle of 130 to 140° and a cell depth of up to 40 µm .

A flat, positive position of the blade improves the ink transfer. The use of a lamella blade is advisable due to the lower flexible pressure. The use of such lamella makes the usual frequent sharpening as for conventional doctor blades, obsolete. These lamellas are available in a variety of different steel qualities, including ceramic surface coatings for extra hardness.

Print down optimization is achieved by adjusting the printing speed, which, ideally, should allow a short flow-out period after ink transfer to the substrate and support optimal plane-parallel orientation of the pigment to the substrate surface.

Printing substrates

The variety of substrates used in gravure printing ranges from strongly absorbent papers used in the wallpaper industry, to specialty papers and plastic films used in the packaging and label industry. The adjustment of color and substrate is basically decisive for the quality of the printing result. The best effects are achieved by selecting substrates with an even and smooth surface. Pigments have the best orientation at such surfaces and can show their optical quality in the best manner.

Further processing and properties

The use of pearl luster pigments in gravure printing has no impact on any additional processing steps, e.g. overprints, top-coats, hot and cold sealing, lamination and embossing.

Final embossing and calendaring steps lead to compression of the applied film, and therefore to thinner film thickness, in combination with an improved orientation of the pigments within the films. This causes finally very brilliant and glossy effects.

The pearlescent luster pigments used are not known to impact negatively on rub fastness and barrier properties, nor on sensory, acid and alkali tests (e.g. wash-removal of labels).

5.3.2 Pearl luster pigments in flexographic printing

Due to their chemical and physical properties, pearl luster pigments are suitable for almost all ink systems implemented in flexoprinting. Regarding quality, the statements made for gravure printing are generally true for pearl luster pigments in flexoprinting.

Printing parameters

Flexographic printing is related to letterpress (relief) printing, so far as they both use relief-printing plates. The raised parts of the printing plate is the printing area, the non-raised parts are the non-printing areas. Flexographic printing-plates are made of molded rubber or flexible photopolymer material, although natural and synthetic rubbers may also be used. They are usually mounted on the cylinders with sticky-back tape. Sleeve technologies are also often used with the advantage of a faster change of the printing plate.

Flexoprinting presses may be single- or multi-cylinder types. The former having a central impression cylinder for two to seven plate cylinders (satellite principle), and the latter a series of plate cylinders arranged one after the other, or banks of cylinders in a stacked arrangement, with one to four cylinders on either side of a vertical frame.

A typical flexoprinting press has four rollers; a rubber inking roller transfers ink from the ink tray to a screen or anilox roller, followed by the plate cylinder and impression cylinder. Flexoprinting is mostly used for packaging and self-adhesive labels.

Printing is normally performed onto rolls (or "webs") of substrate, and the printed rolls are fed directly into the packaging and finishing process.

Printing process

Pearl luster pigments are suitable generally well suitable for low viscosity solvent or waterborne flexo inks. To achieve the optimum pearlescent effects only highly transparent binder systems should be used. To enhance the incorporation and transfer of pearl luster pigments, an ink vehicle with slightly higher solids is beneficial. By tinting the formulation with transparent pre-dispersed organic or inorganic toners, tints or pigments, a variety of brilliant colors can be created. However, since the pearlescent effect may be somewhat compromised by tinting, a good recommendation would be, using only up to 3 wt-% of the conventional pigments, in the ready-to-print formulation.

Satin effects can be achieved by using a particle size range of 5 to 25 µm with a concentration of up to 30 wt-% pearl luster pigments in the ready-to-print formulation. For enhanced brilliance, particle size ranges of 10 to 60 µm and 10 to 125 µm are recommended using slightly lower concentration levels of 10 to 25 wt-% pigment in the ready-to-print formulation. For sparkle effects, large particle sizes of up to 200 µm with a concentration of 5 to 15 wt-% are necessary (e.g. pigments of the "Miraval"-type based on borosilicate platelets).

The surface structure of pearlescent pigments contains extremely fine capillaries. Therefore, pre-wetting helps with the incorporation and the homogeneous dispersion, while improving the anti-settling and adhesion properties.

The dry pigment should be wetted out with components of the ink vehicle, e.g. water, alcohol or any other thinner, compatible with the vehicle. The resulting pigment paste should soak for a few hours or overnight, and may then be added to the ink vehicle. This procedure also reduces dusting during the final incorporation of the pigment into the ink system.

The already mentioned pre-wetted pearl luster pigments in form of pellets in Chapter 5.3.1 are also suitable for flexographic printing.

For roll-to-roll printing on coated paper or films, a slightly higher printing viscosity (40 to 60 s in a DIN-Cup 4 mm cup) is recommended. For solvent-based inks, as used for printing on plastic films the viscosity is usually lower (25 s) than the viscosity for water-based inks. For UV inks the viscosity is relatively high, in the range of 50 to 60 s. These recommendations should, however, only be regarded as general guidelines. Optimal concentration and viscosity are always dependent on the printing speed, printing substrate, and machine configuration and need to be adjusted accordingly.

Photopolymer plates for flexoprinting with pearl luster pigments

The choice of a suitable softer photopolymer plate with a lower surface tension prevents the accumulation of pigments, allowing an optimal transfer of the pigment particles, which is important for the production of a uniform, un-blotched print image. This also prevents the accumulation of pigments on the plate, thus avoiding squeezing lines.

As an example, the following plate types represent suitable printing forms:

- DuPont "Cyrel PLS", "POF", "UVB", "DPS", "DPN", "DPH2", "DPC"
- BASF "Nyloflex FAR", "FA", "FAH", "Gold A"
- Aquaflex

A multilayer plate, which is constructed with an elastic carrier layer, a stabilizing film and a comparable hard photopolymer relief layer, is less suitable for printing pearl luster pigments.

However, the printing experiments regarding the suitability of different polymers of various suppliers, plate thicknesses for the flexoprinting process, and varnishing or coating, is an ongoing theme, as technology is continually changing.

Anilox rollers for flexoprinting

Flexoprinting inks with pearl luster pigments are usually printed on multi-color flexoprinting machines. In addition, they are widely used in chamber type coating units (containing a doctor blade and a ceramic anilox roller) on offset presses. Generally, laser-engraved ceramic rollers are beneficial within the process. Due to their specific cell geometry they allow excellent ink transfer.

The dip or screen rollers (anilox rolls) used in flexoprinting range from traditional steel screen rollers with many different surface structures to sleeves and laser-engraved ceramic rollers. The selection of screen roller depends on the printing substrate, the ink used and the heaviness of the print. The best results with pearlescent pigmented flexo inks are obtained using laser-engraved ceramic rollers, most probably owing to their special cell geometry, which allows very good ink transfer, especially with water-based systems.

The cell:wall (land) proportion of the anilox roller should not be less than 4:1, because the larger the cell wall, the more the pearl luster effect is reduced. In general, the finer the screen, the poorer the pearl luster effect. For certain applications and very high transfer volumes of vehicle and pigment it is advantageous to use so called ART (Anilox Reverse Technology) engraved screen rollers.

Depending on the substrate and the print design, special screen sizes must be used for full-tone and halftone printing (Table 5.1). In flexoprinting optimal effects are achieved by printing onto smooth, even surfaces. However, pearl luster pigments can also be used on relatively rough and absorbent substrates. A flat positive position of the blade can here improve the ink transfer.

Table 5.1: Correlation between pigment particle size, printing substrate, printing image and screen

Pigment particle size	Printing substrate	Printing image	Anilox (screen)
5–15 µm 10–60 µm 5–125 µm 20–200 µm	rough and absorbent material	area (full tone) printing	32–72 lines/cm
5–25 µm 10–60 µm	smooth material, e.g. coated paper and films	area (full tone) printing, half tones and fine detailed printing	60–80 lines/cm
5–15 µm 5–25 µm	smooth material	half tones (screen >60 lines/cm)	80–120 lines/cm

Overview
Flexoprinting inks are low-viscous, solvent-based or water-borne liquid inks; they are fast drying, especially the UV curing systems, which are becoming more and more widespread. The reason is that manufacturers are on the constant lookout for new and improved ways of overcoming environmental issues such as pollution of the environment with solvents, disposal of leftover materials and recycling/disposal of printed products. As well as offering advantages in this respect, UV systems are an excellent choice for turning out high-quality printed products.

The print result is again dependent on the amount of printing ink that is transferred to the substrate. The concentration of pearl luster pigments is approximately 15 to 25 wt-% relative to the ready-to-print ink. Print down can be optimized by adjusting the ink viscosity and printing speed, which, ideally, should allow a short flow-out period after ink transfer to the substrate and support optimal plane-parallel orientation of the pigment to the substrate surface.

Printing substrates
Compared to gravure printing, pearlescent flexo inks can be better used on relatively rough and/or absorbent surfaces (e.g. tissues). Best effects are achieved, however, on substrates with smooth surfaces, which allow for optimum lay down (plane parallel orientation) of the pigments. The substrates of choice, therefore, are satin papers and plastic films and sheeting such as those used in the packaging and label industries.

Further processing and properties
Pearl luster pigments do not affect the further processing of flexographic printing products. Further printing and overprinting, by a different technique if desired, can be handled as usual. Hot and cold sealing and laminations can also be performed in the usual way. Pearl luster pigments used are not known to impact negatively on fastness properties.

5.3.3 Pearl luster pigments in screen printing

Screen printing is a printing technique, where the ink is forced through the openings of a web, by a flexible rubber, or synthetic blade onto a supported substrate. The flatbed printing process uses a frame, covered with the web, which functions as image carrier. A squeegee, running within the frame presses the ink through the web onto the substrate.

The rotary screen printing process uses a hollow cylinder consisting of a metal web, carrying the stencil. A squeegee running inside the cylinder, presses ink onto the substrate. Modern, automatic presses for screen printing have up to eight inking systems in an inline or turntable arrangement (in textile printing even up to 16 inking systems). The screen material and its filament structure must permit smooth ink transfer to the substrate. Both have a decisive effect on print down quality. The function of the screen is to support the stencil as well as to carry the ink.

The fabric of the web may consist of metal or of fine synthetic or natural filaments, such as polyester, polyamide or polyester/polyamide mixed weaves, having either multifilament or monofilament fabrics. The material structure must permit a continued smooth ink transfer onto the substrate.

The filament thickness determines the layer thickness of the ink on the substrate. Both multifilament and monofilament fabrics are in widespread use. The most commonly used screen materials being silk, polyamide, polyester, metal coated polyester, polyester/carbon polyamide, and stainless steel.

Screens may be coarse, i.e. with just a few filaments per cm (e.g. 12) or fine, with a large number of filaments per cm (approximately 200). The mesh number indicates the number of filaments per cm the screen has (e.g. monofilament polyester fabric 77/55; 77 filaments per cm, 55 μm filament thickness).

For pearl luster pigment applications in screen printing, monofilament fabrics are recommended, as these have less of a tendency to clog with pigment particles than multifilament fabrics. Consideration must be given to the appropriate mesh size, depending on the chosen pigment type and size. The mesh size should be 1.5 to 2.5 times larger than the maximum pigment particle size. In the case of a pearl luster pigment with a particle size distribution of 10 to 60 μm, the mesh size should be 60 μm * (1.5 to 2.5) = 90 to 150 μm. Also, screen fabric conditioning processes such as calendaring, as well as the fineness of the fabric filament (former S, M, T or HD, today the exact filament thickness is stated) are important for print down quality. Both parameters influence ink and pigment transfer as well as the volume of ink transferred. The pigment, substrate, ink, fabric type and filigree must all be properly matched to the print job.

Printing parameters

Screen printing inks need to have good application properties such as compatibility, printability and flow characteristics as well as good fastness properties such as light-fastness, scratch resistance and washability. The ink should not clog the screens and must have the correct drying characteristics.

Since screen printing can be carried out on virtually all types of substrate, including paper, carton board, plastics, timber, textiles, metals, glass and ceramics, the variety of binder systems available is enormous. The binder systems of an ink are there for support, cohesion and pigment transfer. They give the ink its printing characteristics and play a major role in ensuring that the ink adheres properly to the substrate. Binders are synthetic or natural resins dissolved, or dispersed in a solvent. Those most commonly used are linseed oil, shellac, epoxy resins, polyurethane, polystyrene, polyvinyl acetate, polyacrylate, melamine-formaldehyde and cellulose acetate.

The binder is important for determining the ink's drying characteristics, particularly its solvent retention. Solvent retention in a binder thus also affects the choice of solvent used in an ink system.

The solvent in turn significantly affects the consistency of the ink. Sometimes it is there for another reason, too: to prime or soften up the surface of the substrate to ensure that the ink will bond. The main solvents used in screen printing inks are: acetone, ethyl acetate, toluene, ethyl alcohol, butyl acetate, xylene, butanol, methyl glycol acetate, cyclohexanone, ethyl glycol and benzyl alcohol.

Modern screen printing inks tend to be solvent-based, whereas those used for textile pigment printing are almost entirely water-borne dispersion inks. It is generally hoped, however, that solvent-based inks will be superseded at some time by water-based or low-solvent systems. There is room here for more widespread use of UV-curing inks, as these have the advantage of being fast drying and solvent-free; also, the wet film exactly resembles the dry film and there is no ink drying in the printing screen.

Pearl luster pigments are nowadays widely used for screen printing. Showing excellent compatibility with most of the binding systems and solvents mentioned here, they allow for a wide range of creative effects. It will be found, however, that the clearer the binder is, the more brilliant will be the impact of the pearlescent pigment. Hiding colors or pigments should not be used, as these mask the pearlescent luster. Colorations with well dispersed, transparent pigments (up to 20 wt-% relating to the pearlescent pigment) considerably extend the effect possibilities. To incorporate the pigment into the ink system it is recommended to pre-wet the dry powder pearlescent pigment. The usual concentration of pearlescent pigments in screen printing inks is approximately 8 to 15 wt-% relative to the ready-to-print ink.

A variety of substrates ranging from paper to ceramics have already been mentioned. Printing pearl luster pigmented screen printing inks requires an optimum choice of binding and coloring systems, and printing technology commensurate to the end product. The choice of screen fabric is also extremely important.

The screen printing process basically allows the use of all pigment particle size ranges. A mesh size 1.5 to 2.5 times the maximum pigment particle size should be chosen. Other considerations such as the choice of squeegee, the printing speed and drying behavior depend on the substrate and the ink used, which must be well matched.

Ink drying in screen printing is a critical parameter based on the relatively large film thickness deposited on the substrate. Owing to the morphology of the pigments, pearl luster pigment inks may take longer to dry (or the drying or fixation temperature may need to be raised slightly) if the product is required to have specific fastness properties. If a multiple-color wet-on-wet print with pearlescent pigments is intended, the use of intermediate drying stations is advantageous, as there may otherwise be carry-over of ink from one printing screen to another. Meanwhile printing ink manufacturers have developed new systems, which allow wet-on-wet printing without intermediate drying steps. Pearl luster pigment ink has tended to be used as a feature ink, following in the last printing station the printing of the process colors.

Printing substrates

Although, as already mentioned, the substrates include paper, carton board, plastics, timber, textiles, metal, glass and ceramics, the most important screen printing application for pearl luster pigments is printing on textile fabrics. Textile pigment printing is mostly done with water-based ink systems. Pearlescent pigments are especially suitable on this type of substrate because of their excellent water- and alcohol-resistance, their thermal fixation properties, and their solvent-, wash-, sweat- and chemical cleaning resistance.

Pearl luster pigments are often used in textile printing inks to create brilliant designs. The pigments can be used either in a transparent ink vehicle or in combination with other transparent dyes or pigments. For pigment particle sizes of 20 to 200 µm a pigment concentration of 0.5 to 5 wt-% and screens of 27 to 32 lines/cm with a mesh opening of 250 to 270 µm are recommended.

For pigment particle sizes of 10 to 60 µm a pigment concentration of 10 to 15 wt-% and screens of 40 to 60 lines/cm with a mesh opening of 90 to 150 µm are recommended. For pigment particle sizes of 10 to 125 µm a pigment concentration of 8 to 10 wt-% and screens of 32 to 44 lines/cm with a mesh opening of 135 to 200 µm are favorable.

Pearl luster pigmented textile inks which are waterborne systems, need to be thermo-fixed to the fabric at a temperature range of 150 to 160 °C for about 5 to 10 min, after pre-drying.

This thermal fixation ensures good ink adhesion, which in turn provides excellent wash and bleach resistance.

The surface structure of the printing substrate is of minor importance in screen printing, so that even highly absorbent materials such as textiles and ceramics are readily printable.

Pearlescent pigment inks for paper applications are mostly solvent-based on acrylics.

For pigment particle sizes of 10 to 60 µm a pigment concentration of 10 to 15 wt-% and screens of 60 to 80 lines/cm with a mesh opening of 75 to 90 µm are considered. For pigment particle sizes of 10 to 100 µm a pigment concentration of 8 to 10 wt-% and screens of 40 to 70 lines/cm with a mesh opening of 90 to 120 µm are advisable.

Pearl luster pigment inks for plastics, film and other non-absorbing materials are mostly solvent-based or UV-radiation curing systems, depending on the substrate. UV inks are fast dried (cured) through the photo-polymerization initiated by ultraviolet radiation, and beneficial for applications on non-absorbing materials such as metal, plastic, and wood.

For pigment particle sizes of 10 to 60 µm a pigment concentration of 10 to 15 wt-% and screens of 40 to 60 lines/cm with a mesh opening of 90 to 150 µm are suitable.

Some ink components and any stray UV-radiation from the drying system can be detrimental to ones health. It is therefore essential that appropriate safety measures are taken with regard to eye and skin protection. The covering power (opacity) is not always perfect, as the ink layer has to be thin enough to get polymerized and dry. For UV-curing screen inks special formulations are recommended as they are easy to print and exhibit excellent luster.

Further processing and properties

Pearl luster pigments do not negatively affect follow-on processes. Further printing steps, overprints, etc., even by different printing processes, can be carried out as usual, as can hot and cold sealing, and lamination. Also, the pigments have not been found to reduce fastness.

Superb enhanced effects can be achieved by calendaring, a process which is particularly apt in textile printing for the achievement of chintz effects. The excellent chemical resistance of the pigments means that other methods such as wash-out/stone-wash applications for denim or alkali creping applications can also be applied.

5.3.4 Pearl luster pigments in wet offset printing

Modern lithographic printing utilizes thin photomechanical treated aluminum printing plates and dampening solutions to separate the printing from the nonprinting areas. The plates are mounted around the circumference of the printing cylinder and, because they are easily worn, the image is not transferred directly to the substrate but first to a rubber blanket roller, hence the name offset. The inking system is the same as that used in letterpress, except that it also incorporates a dampening system. Originally, offset printing was done with sheet presses.

The forte of offset printing lies in its fine image transfer to variously structured substrates at high speeds. It is a versatile process suitable for long runs of newsletters, packaging, etc. in sheet offset (onto paper, carton board, tinplate) and, increasingly, for newspapers in web offset.

Inks and inking system

The purpose of the inking system is to apply a steady supply of ink to the lithographic plate. Its design greatly affects the quality of the print. It comprises an ink tray and a system of rolls for applying ink to the plate.

In offset printing the printing ink is transferred to the substrate by a plate having planar printing and nonprinting areas. Inking of the printing parts of the plate is based on the incompatibility of oil and water. The surface of an offset plate always has two components:

- hydrophilic (water-attracting) material for the nonprinting areas, and
- hydrophobic (water-repelling) surface for the printing areas.

Copying the image onto the printing plate is done by modern lithographic methods, using positive or negative originals. Image transfer must be such that an exact copy of the original appears on the plate. The process must be reproducible to enable duplicate plates to be made from the same film without any variation in image quality. The original must therefore be perfect in every respect, which generally calls for photo retouching. All of the graphic elements must be black and the light-transmitting parts of the film must not show any signs of haze. All artwork must be properly proofed prior to the plate making process. To obtain a faithful copy of the original on the printing plate, the film must always be placed with its coated side next to the light-sensitive side of the plate. The plate is exposed through the film in a vacuum to maintain intimate contact between the film and the plate, so as to prevent the ingress of light from the side.

Uncovered areas of film, i.e. those that are exposed to light, undergo a chemical change which results in the sensitized layer being dissolved away. Parts of the plate covered by the black film are unaffected. Most plates are now pre-coated and sensitized, being supplied in light-impervious packaging. Following exposure, the plate is developed with water or a weak alkali, depending on the copying process. Plates are developed in numerous ways, either manually or by machine. With negative plates it is the unexposed areas that are developed away; with positive plates the exposed areas are removed.

On nonprinting areas this leaves the hydrophilic aluminum plate material uncovered, while the image-carrying areas retain the thin surface coating. This coating is thus the ink transfer medium.

Today most printing plates are generated by CTP (Computer to Plate) technology. The advantage for the user is clearly in the reduction of use of chemicals, rapid data transfer, and of course the saving of the classical reproduction and retouching processes.

Fixing the plate or treating it with dilute phosphoric acid enhances the hydrophilic properties of the image-free areas. Finally, the plate receives a thin film of gum to support hydrophilic property and protect it against oxidation. Being hydrophilic, the non-image carrying areas of an offset plate are ink-repellent. During a print run they are subject to mechanical and chemical wear. Substances are added to the dampening solution to stop the ink repellence from waning. These so called fountain additives usually contain buffer systems to regulate the pH value.

Field tests have shown that, with careful matching, standard dampening solutions can be used provided they are matched to the pearlescent pigment type used. However, the pH value of the fountain solution should be kept to 4.5 or below, or the use of a buffered solution, to avoid any drying problems, and pigment migration to the dampening solution.

Common offset printing inks are compact and highly viscous, the same as in relief printing. They are soft and tacky, enabling them to be applied to any dry surface. Like any printing ink, they have two components: the pigment and the binding system which, in offset printing inks, consists of a colloidal solution of synthetic and/or natural resins and drying oils in mineral oil. Print quality depends on the composition of the ink and the way in which it was manufactured. The binding system is important for transporting the pigment particles and affixing them to the paper or other substrate during printing. The pigments, on the other hand, provide the coloration. The right pigments must be selected to suit the composition of the ink. The pearlescent pigments used in offset printing must be specially designed and selected for the job. Selection is appropriately carried out in consultation with the ink manufacturers. For optimum ink transfer to occur during printing, it is important for the ink to contain not only the pigments but also special binders with the right physical-chemical properties. There are ink manufacturers who supply these products in ready-to-print form.

Suitable substrates are the same as those mentioned for gravure- and flexographic printing: all packaging and label materials and ordinary print-job substrates. The substrate surfaces should be as even and smooth as possible to allow for optimal lay down of the pearl luster pigments. For ideal effects to be achieved this is especially important, as the ink layer in offset printing is relatively thin.

The use of pearl luster pigments in offset printing will require attention to the following points. For best print results in terms of ink transfer, print-run behavior and visual effects, it is mandatory to use only ink products from quality ink manufacturers.

Dampening system

Having printing and non-printing areas horizontally on the same plane level requires not only a steady feed of ink but also constant replenishment of water lost through printing and evaporation. The dampening system of an offset press fulfils this function, applying a thin film of dampening solution to the plate. The dampening system is directly ahead of the inking system and most commonly consists of a water tray with a system of application rolls.

Pearl luster pigments for offset printing

A special non-dusting pigment blend has been developed known by the trade name “Iriodin Pearlprint” Litho S, or the highly viscous pigment paste “Biflair” 66A, which can be dispersed in regular transparent offset binders (transparent white) and varnishes [5].

The offset printing inks set up with the mentioned pearl luster pigment preparations may require the adjustment of the ink viscosity and tack depending on the press conditions and the substrate. They can be used up to 35 wt-% in relation to the ready-to-print ink. Due to the composition of these pigment preparations, increased ink feed will be required to allow the pigment particles to pass the roller nips. The most brilliant effects in offset printing with pearl luster pigments are achieved by using full tones. Halftones are achieved by varying the dot size reduce transfer and therefore compromises the luster of the print.

The strength of the pearlescent effect is dependent upon the color sequence. Pearl luster pigmented inks can be run either first down or last down in the offset printing process. Naturally, ink in the first print station should have a higher tack compared to ink running in the last print station. The optimum pearl effect would be achieved by printing pearl luster pigments full tone at the last print station on top of the four-color process. To create a variety of colors with pearl luster pigments, only highly dispersed, transparent organic or inorganic pigments should be used for shading. However, since this shading compromises the pearl luster effect, tinting only up to 3 wt-% with other inks in relation to the ready-to-print ink is recommended.

Pearl luster pigments in the printing process

For the best ink transfer, blankets that provide a ground-polished surface and a low-stretch/higher hardness carcass should be used. A smooth blanket surface is beneficial to avoid the piling of pigments on blanket and plate. The new developed pearl luster pigment series can be printed with standard fountain solution.

Printing plates should be post baked for extra hardness which is advantageous for long run printing. The post baking process is commonly used for conventional plates as well as for digital plates.

The substrate surface should be as straight and smooth as possible to allow for optimal orientation of the pigments. When working with a mat-coated paper, the outcome will be a shiny, brilliant luster; when working with a glossy paper, the outcome will be a velvety, satin effect.

Further processing and properties

The use of the new pigment preparations has no impact on the further processing steps e.g. overprints and topcoats, hot and cold sealing, lamination and embossing.

No changes with respect to scratch resistance, barrier properties, sensorial and acid/alkaline resistance were encountered.

5.3.5 Pearl luster pigments in other lithographic offset printing methods

Meanwhile pearl luster pigments are applied also in other offset printing processes. Waterless printing is a type of offset printing that does not require the use of dampening solutions. The various waterless techniques include Toray-waterless and letterset. Pearlescent pigments are also used in the classic letterpress.

Characteristic features of letterpress are its clear, sharp image and intense coloration. This advantage over wet offset printing is also seen with pearl luster pigments and is explained by the fact that the applied ink layer is thicker than in wet offset. Secondly, by the absence of a dampening solution that inevitably initiates emulsification and tends to give the print a somewhat matter appearance.

The technique is used to print newspapers, magazines, calendars, greeting cards, mail pieces, cartons, and display boxes.

Printing inks

Toray printing plates require the use of special Toray-inks from manufacturers who have specialized in inks for this particular technology. Developments in this area have produced spectacular results, the effect achieved being superior to that of wet offset. The use of the earlier mentioned pigment preparations for the manufacturing of those inks is recommended.

One important point, however, is that presses designed for use with Toray-plates need special print cooling, as without it only short runs are possible without the plate beginning to tint. Toray-inks can only be used in quite specific temperature ranges.

Letterpress inks are usually highly viscous pastes that may be extended with oils, glycols or water. The same principles apply to the use of pearl luster pigments as already outlined in Chapter 5.3.4. in respect to printing process, substrates, and further processing/properties.

5.3.6 Pearl luster pigments in offset coating

Offset coating is an additional possibility for refining of printed products when using pearl luster pigments. There are two main applications used for pearl luster pigments in offset coating. Here are varnishes applied in thin layers to partial or solid areas. The function of the varnishes is to protect previously applied coloring and decorative finishes or to give a value-added effect (high luster or special surface finish).

The process is based on either water-borne acrylic dispersion lacquers, or UV-curing systems which are applied by using specially designed coating units (chamber systems) attached to modern offset presses. The coatings may be applied full-area or selectively. These varnishing towers are fitted with flexo-plates and screen/anilox-rolls. These presses open up new ways of adding value to printed products through the use of pearl luster pigments.

Pearl luster pigments are a considerable enhancement for the use in both coating types. They may contain up to 25 wt-% pearl luster pigment, depending on the effect that is desired. Preparing pearl pigment varnishes follows the recommendation, given before in the chapter of flexo printing (Chapter 5.3.2).

The use of "Iriodin" Pearlets for these applications is recommended [5]. "Iriodin" Pearlets are a new type of pigment preparation for simple, problem-free preparation of effect printing inks and inline coating systems/overprint varnishes. They offer significant savings in time for ink manufacturing due to the better wetting of the pigments.

These Pearlets are superior to conventional powder pigments in a number of ways. They have a smaller volume, ease of handling, dust-free operation, precise metering, better dispersion behavior, optimized sedimentation and re-dispersion properties.

They were specially developed for use in flexo and gravure printing, in water-based, and UV-over-print varnishes (in conjunction with a chambered doctor blade system).

"Iriodin" Pearlets are available in 4 versions, or types:

Type W: for water-based ink systems,

Type L: for solvent-based ink systems,

Type UV: for UV curing ink systems,

Type P: for paper sizing and special coating applications.

Depending on type, these pigments contain between 1 and 4 wt-% and/or 12 and 16 wt-% polymer in the formulation – including moisture retardants. This means, assuming 20 wt-% pigmentation, there is max. 4 wt-% resin added to an ready to print system.

5.3.7 Pearl luster pigments in the bronzing process

Bronzing is an old-established method used to produce eye-catching effects. It is mainly used in packaging and label manufacture.

In the bronzing process, pigment powder is dusted onto paper or carton sheets holding a previously offset printed tack, sometimes a colored one. The pigment powder is applied by using a plush covered roller, which is fed by a scoop or dip roller, placed above the application roller. All area of the paper or carton sheet, not covered by the previously applied tack, will not attract and not hold pigment particles.

Coarse adjustment of the bronze flow is done by opening and closing the flaps of the application unit, according to the size of image area. Fine tuning of pigment flow is performed by adjusting the speed of the application roller, and the scoop roller.

Once the required amount of bronzing powder has been applied to the areas to be bronzed, these are burnished with wide, plush-covered burnishing pads. The pads apply an even pressure over the entire printing area and vigorously rub the pigment powder into the under color.

Plush-covered belts remove excess powder from the image-free areas. The belts move in opposite directions and sweep the excess pigment into the extraction system below. After leaving the bronzing station, the bronzed and partially de-dusted sheet enters the cleaning station. Here the sheet is further burnished and de-dusted. In the final step, the sheet runs through two top and one bottom plush-covered roller, to emerge clean top and bottom. The lifting of the sheet off the transport web is done by a bristled roller which contacts the web. The bronzing, and the cleaning is finalized and the sheets are stacked. This established technique is used until today with good success.

For use in the bronzing process, a range of special pearl luster pigments was developed, which are tailored to the technical requirements of the bronzing process. These pigments are available under the trade name "Iriobronze" with particle sizes ranging from 10 to 200 μm.

The advantages of bronzing are in the high-gloss and high-luster effects, which can be achieved due to the relatively large size of the pigment particles. The disadvantage of the bronzing technique, is associated with dusting, although suitable extraction systems will keep this to a minimum. Bronzed surfaces tend to be rather rough, have poor scratch resistance, and need varnishing. Machine bronzing of thin papers is sometimes problematic depending on the pigment concentration, and running speeds are affected by the size of the surface to be bronzed. Occasionally, longish drying periods are required to enable the pigments to become firmly incorporated into the tack.

Printing parameters

To obtain optimal results, the bronzing machine must be suitably set up for the material to be bronzed. The de-dusting plush pads and belts must be in proper condition, enabling smooth removal of excess pigment. Old and extensively used plush pads and belts will lead to scratch marks within the image area. Tack should be generously applied by offset printing. If threadlike structures are seen at the end of the process, the amount of tack is correct and the "Iriobronze" will be wipe-fast once incorporated. The amount of under color required is about 2.0 to 2.5 g/m^2.

The dampening solution should be reduced to a necessary minimum. The use of additives should be avoided, except for alcohol dampening systems. Normal drying takes about 24 hours and, after 48 hours, the pigment should adhere perfectly to the surface. Re-stacking may speed up drying by increasing oxygen exposure. Using bronzing additive (e.g. R6) may be of advantage for facilitating and improving the cleaning of prints without negatively affecting brilliance or luster effects. To achieve optimal results, adding 3 to 5 wt-% relative to the pigment is adequate.

Printing substrates

Bronzing is primarily used for label and packaging manufacture, usually paper for wet label application within the normal weights of 70 to 120 g/m². Carton stock is used for candy and cosmetic packaging. Excellent results are also achieved with cover papers or films used in the manufacture of chocolate-box inlay cushions. In general bronzing is suitable for any material that can be pre-printed with high viscous tack by offset or letterpress/letterset.

There is a discernible trend towards the use of other printing methods (e.g. gravure or flexo-graphic printing) to simulate the same results as those obtained by bronzing. This is probably due to the previously described disadvantages of the process, the relatively low production speed and hence also relatively high costs.

Further processing and properties

"Iriobronze" pearl luster pigments can be used in bronzing applications without negatively influencing further processing steps. Further printing, overprinting/lacquering, hot and cold sealing, embossing, calendaring and lamination can be carried out as usual. Embossing or calendaring even improves the effect, due to optimal orientation of the pigments.

The application of top coat finishes, spirit or nitro lacquers for top coating improves scratch resistance and may enhance luster.

5.3.8 Pearl luster pigments in other coating processes

One special process worthy of mention, however, is coating of paper and fine carton board with pearlescent pigments, as surface finishes applied by this method produce startling effects owing to the thickness of the coating. Many processes can be used to apply the coating, including knife-and-roll, and air-knife-coating.

Coating machines tend to be large in size and coat onto web substrates. They must be capable to apply high volume of vehicle, either by applying several coatings one after the other or by applying a single thick layer.

Coatings are highly loaded (with up to 40 wt-% pearl luster pigment) and are applied in weights of 8 g/m² and more (wet coat). The trend now-a-days is towards the use of water-borne systems. For the further processing of such coated products, papers and carton, see the statements made within the chapters gravure printing and flexo printing.

5.4 Evaluation of the print results

Essentially, the evaluation of pearl luster pigment print results is done visually, attention being given to characteristics such as pearl effect, satin appearance, brilliance and toning. In addition, fastness properties of all types, such as light fastness, scratch resistance, acid and alkali resistance, adhesion, and amenability to other processes such as lamination and embossing are further criteria to be considered.

5.4.1 Print down and print run characteristics

Pearl luster printing inks, coating vehicles, or varnishes have to match the same quality as all other conventional inks and coating varnishes in relation to a print out quality and print run characteristics. A decisive evaluation criteria for the print out quality is the even dispersion of pearl pigments within the printed, or otherwise applied film. An optimal orientation of the flake like pigments, plan parallel to the surface of the substrate is imperative. Only then the characteristic pearl lustrous effect can be achieved and observed.

The optimal pearl effect will be reduced if the pigment is not homogeneously dispersed in the ink, very often caused by an insufficient wetting-out of the pigment with the binder system or carrier (solvent).

If irregularities such as rough print surface, orange peel structure, pigment agglomeration, or show through of the substrate appear, so is this implementing a wrong choice of substrate, not matching printing forms (e.g. screen size is not matching pigment particle size), insufficient ink film splitting, or non optimized printing viscosity, print speed, or drying conditions.

Insufficient pigment wetting and therefore poor dispersion of the pigments in the vehicle, will lead to failure within the print run. Pigment agglomeration, wrongly very often determined as "over-(size)-corn", lead to streaks in the print out, piling of pigment on the doctor blade or lead to sieve clogging.

To eliminate such problems in advance it is recommended to get broad information concerning properties of pigments from the supplier, or pigment manufacturer. Close cooperation with the ink manufacturer is of course essential, and a close collaboration between ink manufacturer, pigment supplier and converter is considered as an ideal condition to develop new area of application, markets and effects.

5.4.2 Coloristics

A coloristical evaluation of pearl luster pigment proofs is very difficult. With the CIELAB color system evaluation is only possible with large layer thicknesses on defined substrates.

Pearl luster pigments, thanks to their transparency, light refraction and scattering properties, can be used to advantage in association with other toning pigments, under colors and colored overprints to create almost endless color variations. Specialty colors, developed through empirical color mixing series, are reproducible only by exactly maintaining the given parameters of the color formulation and application.

The use of color densitometers should be regarded only as a rough aid to coloristic evaluations. The final judgment is done visually by checking the luster and color effect at two viewing angles and then comparing these parameters against the original. As an aid, the I.C.C ("Iriodin" Color Compendium) depicts a wide range of color and effect possibilities [2].

References

[1] Böhm, K., Pfaff, G., Weitzel, J., "Ein effektvolles Farbenspiel - Wie Perlglanzpigmente in Druckfarben für verschiedene Druckverfahren eingesetzt werden können", Farbe + Lack, **105** (10/1999) 30

[2] "Iriodin" Colour Compendium, E. Merck, Darmstadt, 1993

[3] "Effect pigments for the printing industry", Technical information brochure, Merck KGaA, Darmstadt, 2003

[4] "Pocket Pal", 14th Edition, International Paper Company, Memphis, Tennessee/USA, 1989

[5] Pfaff, G., Rathschlag, T., "New Design Possibilities with Special Effect Pigments in Printing Inks", Paint India (4/2002) 65

6 Special effect pigments in cosmetics

Dr. Gerhard Pfaff

Cosmetics have been used for thousands of years. However, since the 1960s, the variety of colors available has grown enormously. In general, the cosmetic industry is very directed on fashion and affected by the wish for something eye-catching and new. In color cosmetics, effect pigments and in particular pearlescent pigments play an important role, not only for their luster effect, but also for color [1–4]. In addition to the water-soluble dyes and organic and inorganic pigments that are used in cosmetics, pearlescent pigments and the related effect pigments, both referred to below as pearlescent pigments, have established an important place for themselves in the market.

The special luster and color characteristics of pearlescent pigments give the color palette of lipsticks, eye shadow, nail varnish and other cosmetics a constant stream of new and attractive trends and effects. But the use of pearlescent pigments is not only restricted to decorative cosmetics. They are also used in body-care products such as shampoos, bath lotions, hair-care products, body lotions and skin creams to improve their product attractiveness.

The concentration used depends on the formulation and the desired effect. For example, a shower gel with only 0.05 % pigment content achieves an impressive sheen or glitter effect without leaving a visible trace on the skin after use. The higher the level of pearlescent pigment, the stronger the visible effect on the skin. This is particularly apparent in decorative cosmetics. Colorful eye shadow powders sometimes contain as much as 50 % pearlescent pigment and thus have a strong visual effect and covering power.

The platelet shape of pearlescent pigments offers advantages over and above the optical color effect. They exhibit smoothness and softness during application. Furthermore, as well as adding color, they improve the homogeneity of application of lipstick and eye shadow powders. Unlike conventional colored pigments, effect pigments do not need to be de-agglomerated prior to incorporation. They can be added to emulsions, pastes, powder mixtures and solutions without an additional homogenizing step.

If a silky shimmer is required, particle sizes of up to 60 μm are used. On the other hand, for a stronger sparkle effect, a size of up to 250 μm is needed. In particular the recently-developed calcium-aluminum-borosilicates (glass flakes) and silica flakes have brought immense variety to the color palette of cosmetics (see Chapters 2.10 and 2.11). The brilliant surface gloss of borosilicate flakes and the purity of the substrate color give lipsticks, nail varnish and other cosmetics eye-catching glitter effects that are not possible with mica-based pigments. Appropriate combinations of color travel pigments (based on SiO_2) with "background colors" can be used to give up to four different color nuances. Depending on the product type, soluble dyes, inorganic or organic pigments are used to give the background color. The interplay of interference and absorption colors lead to very interesting, optical "multi-color effects" that are still visible after application.

Effect pigments are mostly used in cosmetics in combination with dyes, inorganic or organic pigments depending on the type of product. Often 10 to 15 different colorants are used for one color nuance. The possible pigment combinations are almost endless. Thus, the cosmetic industry will be able to continue to offer an ever growing range of varieties of color.

References

[1] *Thurn-Müller, A.*, *Hollenberg, J.*, *Liston, J.*, "Special Effect Pigments and Fillers for Cosmetics", Kontakte (Darmstadt) (2) (1992) 35

[2] *Petsitis, X.*, Kosmetik-Guide „Was ist drin, was ist dran?“, vgs Verlag, Köln, 2002

[3] *Fey, I.*, *Petsitis, X.*, Wörterbuch der Kosmetik, 5. Aufl., Wissenschaftliche Verlagsgesellschaft, Stuttgart, 2004

[4] *Petsitis, X.*, *Kipper, K.*, Dekorative Kosmetik und Gesichtspflege, Wissenschaftliche Verlagsgesellschaft, Stuttgart, 2005

7 Colorimetry of special effect pigments

Dr. Peter Gabel

7.1 General information

The observer or consumer of beautiful color effects experiences a subjective sensation of pleasure. This subjective sensation, however, is not adequate for the manufactures and users of pearl luster pigments who have to keep to well defined shade tolerance limits. This is where colorimetry plays a role to permit the availability of objectively measured data of color.

Due to its chemical constitution and physical properties, every single pigment particle behaves like an interference filter. It separates incident light into a reflected portion of one color and a portion of transmitted light with a complimentary color. The important interference property connected to this is the color change at different angles of view.

The colorimetric analysis of effect pigments also varies with the angle of measurement. In addition to this, several aspects need to be considered for the colorimetry of effect pigments [1, 2]:

- The laws of interference for effect pigments are not yet described by a colorimetric system.
- This is especially true for pearl luster pigments, which do not have the theoretically expected absolutely smooth surface.
- The individual pigment particles may be of non-uniform thickness.
- There is diffused light refraction at the pigment edges.
- A parallel alignment of the pigment platelets is necessary for a truly valid measurement.
- Suitable measuring devices, i.e. colorimeters with variable measurement geometry must be available.

All of these points are a challenge for colorimetric analysts; who demand an intensive and critical judgment. But these points still allow different possibilities for an objective measurement of the colorimetric assessment of effect pigments. Such objectivity is necessary for various reasons, including:

- It is the prerequisite for reproducibility in the manufacture of effect pigments within defined standards.

- It defines agreed standards and tolerances between supplier and customer.
- It builds the foundation for the colorimetric assessment and reproduction of color shades that may be formulated with effect pigments.

7.2 Practical color measurement of effect pigments

The parallel alignment of the pigment platelets is an important prerequisite for the colorimetric characterization of interference colors. This can be achieved by dispersing the pigments in an appropriate varnish and spraying onto a prepared panel or by blade-spreading upon a black/white-drawdown card. In addition, it is vital to obtain a pre-defined film thickness.

The following method is very suitable for use in routine measurements [3]: The pigment – with an optimized concentration (14 wt-% pigment to binder) – is dispersed in a nitrocellulose acrylate lacquer and automatically spread using a doctor blade system onto a drawdown card to form a wet film layer 500 µm thick. This sample preparation technique has proven to be highly reproducible and therefore very suitable for routine measurements.

To determine the color definition according to the DIN 6774 standard, a multi-angle color measurement system with direct illumination is used. The data collected under these conditions, L*a*b*-values according to the CIE system (DIN 5033), usually agree well with the visual impression. The results on the black background, depending on the aspecular angle, give information about masstone, scattering and interference color. On the white background the assessed color values are complimentary to the corresponding interference color. The L*-value from this measurement, combined with the L*-value from the black background, establish the relationship for the hiding power.

7.3 Color measurement systems for effect pigments

Color measurement and color management will increasingly play a key role for quality control in the paint industry in the next few years. This is because paints approved by the supplier's instrumental color data are becoming a more important competitive cost factor. The exchange and transfer of data between suppliers and customers is also not only a question of information, but also a question of corporate culture, trust, and mutual respect.

7.3.1 Angle-dependent color measurement – a historical review

For more than 15 years national and international standardization committees have discussed the question under which angles metallic or effect pigments should be measured to assess the light/dark flop and the color flop. At the start of the year 2000 standardization committees such as (or including) DIN (DIN 6157-2) or ASTM (E2194) published testing procedures. This long discussion period may have led to the development of the goniospectrophotometer using differing concepts with variable measurement angles. Despite the fact that a profound step was illustrated in 1974 with the "Trilac" from LERES [4]. For a long time, the modified three-field filter device according to *Hofmeister* et al. [1] from "Hunterlab" was one of the few instruments available. This concept was adapted by *Rösler* [2] with what was formerly known as the "Johne and Reilhofer ER 10".

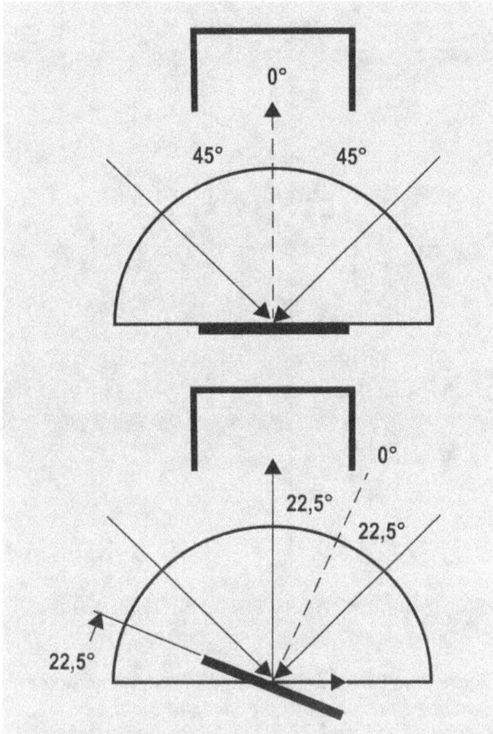

Figure 7.1: Schematic description of the rotatable sample holder to get the angle of half 45°/0° → 22.5°/22.5°

For determination of color travel the measurement system, in this case a "Hunterlab-D25 D 9" tristimulus colorimeter is equipped with an adjustable sample holder. It is then possible to measure color cards or coated sheets at different angles. By turning the sample, the incident angle of 45° can be halved; one obtains a measurement geometry of 22.5°/22.5° and fulfills the requirements for mirror reflection. At the so-called gloss angle, effect pigments demonstrate an optimum of brightness and color (Figure 7.1). The color difference data collected under these conditions, L*a*b*-data according to CIE (DIN 5033), usually agree well to the visual impression and mostly recorded in an a*-b*- and L*-diagram.

The color coordinates (CIELAB-system) of interference pigments are a function of the optical layer thickness on a color position spiral. With increasing layer thickness (e.g. TiO_2-layer), the color coordinates move out of the achromatic range and traverse the range yellow-red-violet-blue-green. In Figure 7.2 (page 200) it can be seen that the powder colors (measurement geometry 45°/0°) of all interference pigments lie in the yellow range and thus all have the same mass tone. The color coordinates of the color cards (aligned particles), on the other hand, are much more differentiated. This is even more so from the measurements obtained under gloss conditions (22.5°/22.5°). As mentioned, an optimum of brightness and color can be seen here.

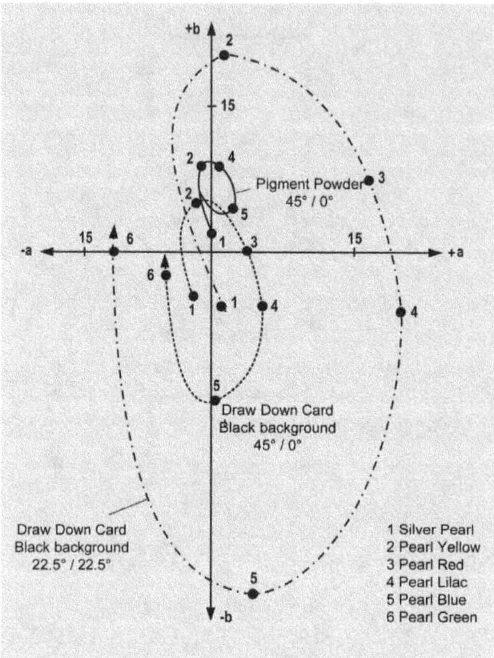

Figure 7.2: Color location spirals of interference pigments in relation to the measurement geometries 45°/0° and 22.5°/22.5° for non-orientated particles (powder) and orientated particles (draw-down card)

The depth effect of a pearl luster pigment coating results because the transmitted light portion meets further layers within the coating, thus achieving multiple refraction. The technological precise assessment of such a luster effect is not possible using conventional luster measuring devices since the aligned reflection only takes place at the surface.

The former "Johne and Reilhofer ER 10" measurement concept "45°/0° with tilt" provides a more comprehensive comparison, especially when measuring interference colors. The sample is measured at various angles of tilt, which allows an unambiguous characterization of the spatial distribution (indicatrix of scattering) of the light reflected by the test specimen [5]. The sample is illuminated by non-directed light (Figure 7.3).

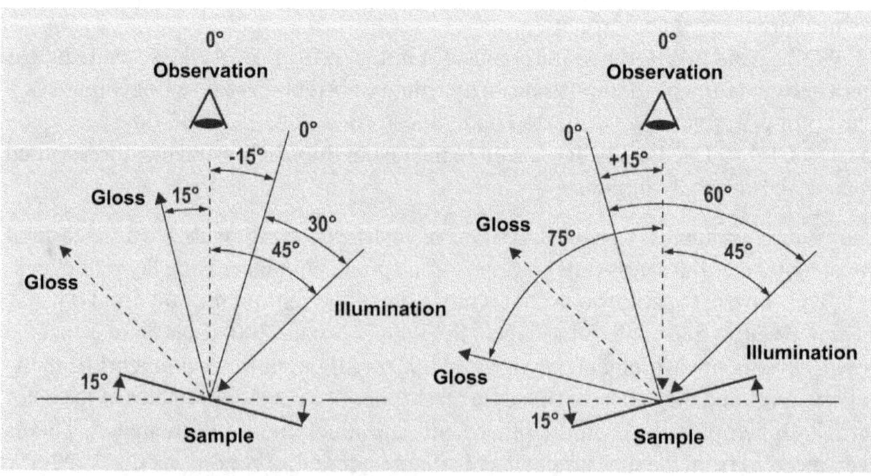

Figure 7.3: Measurement geometry 45°/0° with tilt

7.3.2 Angle-dependent color measurement – a general review

In addition to the previously mentioned measurement systems [6–13] the instruments to be considered now are based upon three completely differing measuring principles:

a) fixed illumination angle of 45°/0° with variable observation angles
b) fixed observation angle of 45°/0° with variable illumination angles
c) variable illumination and observation angles

The color measurement geometries for illumination and observation mentioned in a) and b) are well described in testing procedures such as DIN 6175-2 and ASTM 2194. Description for c) is possible (Chapter 7.3.3) however, there is no agreement at method agreed under any published international standard.

A new proposal by the CIE was made in Lyon 2005. This definition of color measurement geometries is unambiguous and provides a better comprehension and more easily understanding [7].

- Illumination angle : as (aspecular observation angle)
- Examples: 45°: as +15°; 45°: as −15°; 45°: as +110°

The term aspecular observation angle is also commonly known as "degree from gloss", "cis-trans position" or "effect angle" [14].

With respect to method a), various companies produce angle-dependent spectrophotometers. The "Gretag-Macbeth Auto-Eye 642" illuminates the surface from an angle of 45° and offers viewing angles of 25°, 45°, 75° and 110° (aspecular angles). The "X-Rite MA68II", also illuminates at 45°; in this instrument the viewing angles

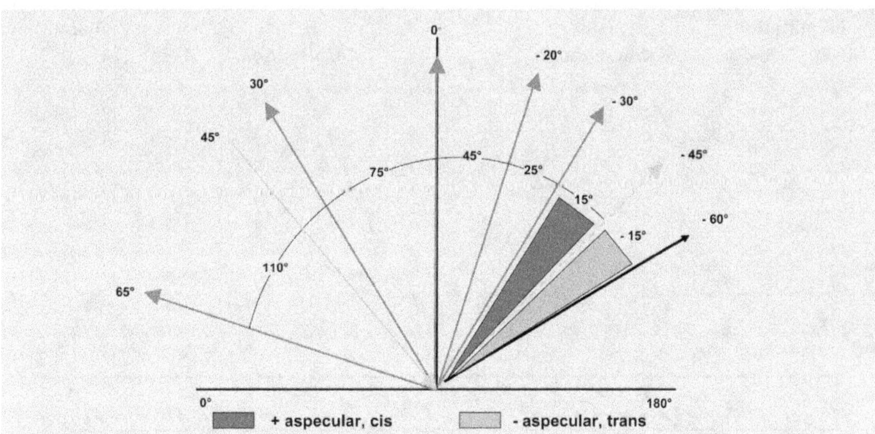

Figure 7.4: Measurement geometries with fixed illumination 45°/0° and variable observation and illustration of the principle of + aspecular/– aspecular or cis/trans-geometries

are 15°, 25°, 45°, 75° and 110° from gloss. With a fixed angle of illumination and optional viewing angles, the two instruments are very similar (Figure 7.4).

However, with respect to method b), the "Minolta" and the former "Optronik" spectrophotometers, illuminate at various angles and measure at a single, fixed angle. With an angle of observation of 0° (perpendicular), the "Minolta CM-512m3" measures at 25°, 45° and 75° under circular illumination. The "Optronik Mulit-Flash", which is no longer commercially available, measured at 45° illumination in eight geometries ranging from 25° to 115°. The effect angle in the latter case changes from the trans to the cis position within the geometric series; i.e. from the perpendicular (+ aspecular) to the horizontal (– aspecular). This change is described as "cis-trans position" [14]. The same effect angle (aspecular angle), i.e. 15° could have two different measurement positions, +15° or –15°. At least the measurement system gives the option for results at the two different positions. The "Phyma WICO 5&5" system measures under illumination at 22.5° and 45°, offering 67.5°, 0°, –22.5°, –45° and –67.5° viewing angles and a large measuring spot. The five presets between 0° and 180° (relative to the horizontal) provide virtually nearly (should be one or the other) ideal pigment definition.

With respect to method c), the "Murakami GCMS-3B" provides complete definition in 1° steps. The instrument features horizontal illumination from 16° to 180° and observation up to 196° (transmission). At this time from the point of view of ASTM or DIN standards there is no agreement on how to describe the illumination and observation angles. The protractor method in Figure 7.5 has no agreed consensus. Totally variable reflectance and transmittance measurements from 0° to 360° can be

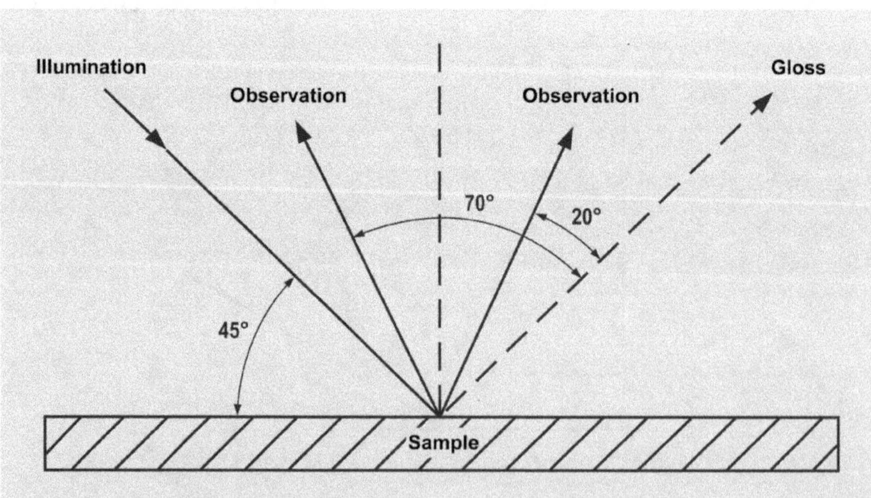

Figure 7.5: Measurement geometries with variable angles of illumination and observation

achieved with the “GON 360” from “Instrument Systems”. The “Zeiss GK 311/M” multi-angle instrument is no longer manufactured, but it was possible to set both the illumination angle and the viewing angle independently of one another in 5° steps. The minimum illumination angle was 25° (horizontal) and the maximum viewing angle 155°.

Comparing results from different multi-angle measurement systems shows, that the instrumental data of the current systems agree very well with the visual shift in first generation effect pigments, but not in second generation materials with their high color shift. As a result, in the DIN 6157-2 standard this class of pigments is even excluded. Today only a few manufacturers offer multi-angle spectrometers. Mostly such systems are useful for R&D only. In the production environment there is also a need for portable instruments with more measuring geometries than those presently available. As an interim step, in existing systems with 45° illumination it might be possible to introduce a –15° aspecular measuring angle in addition to the well established +15° aspecular angle.

7.3.3 Angle-dependent color measurement – new developments

A main feature of effect pigments is the dependence of interference color on the angle of illumination. According to the Fresnel equation for light reflection, the reflection maximum shifts with an increasing illumination angle towards shorter wavelengths with an additional increase in the reflection value. At a fixed layer thickness the incident angle is of utmost importance, because the interference color is dependent upon the phase difference of the reflected light rays.

The obvious assumption that an increasingly flatter angle of incident light makes the higher layer thickness colors visible, matching the longer path, is not valid. Comparative measurements on various pearl luster pigments with spectral photometers have proven this. The variability in results from the methods of measurement mentioned above might be explained by this fact [15–17]. For this reason it is extremely important to establish exact agreement between customer and supplier regarding the measurement angles and geometries in product analysis and specification.

Color measurements produced by varying the horizontal illumination angle (protractor method) and keeping the aspecular angle constant (i.e., 15°) can be portrayed two-dimensionally in the a*b* part of the L*a*b*-color diagram by plotting interference shift (“interference line”) [18]. This interference line is typically dedicated to interference pigments (Figure 7.6). Laboratory measurements stepwise 5° from the horizontal give a very accurate shape of the interference line. However, the amount of data is too great for practical purposes. For quality control and product characterization the reduction of data to the three horizontal illumination angle (protractor convention) – flat (25°), classical (45°) and steep (75°) is sufficient.

The a* b*-diagram in Figure 7.7 points out the results from measurement geometries where illumination and observation is on the same side with respect to the gloss angle (+ aspecular). In this figure the protractor convention is used; i.e. if the illumination angle is 45°, the specular angle is 135°. Two aspecular angles +15° and –15° belong to this specular angle, which lead to the observation angles 120° and 150°. Both measurement geometries give different remission – and therefore different L*a*b*-values.

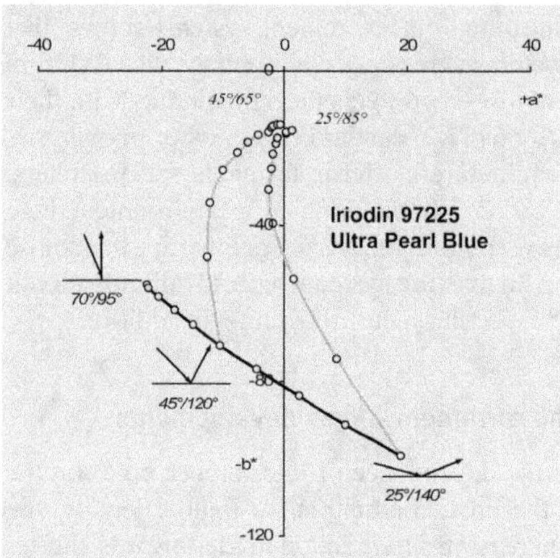

*Figure 7.6: Interference and aspecular shifts of an interference blue pigment in the a*b*-diagram*

A new system developed by Merck and now distributed by Datacolor-ETA-System as the "FX 10" sets out to meet the new requirements: variable illumination angles, classical aspecular angle, +– aspecular angles and a reduced number of geometries [19]. This new measurement tool addresses all requirement and aspects of current effect pigments:

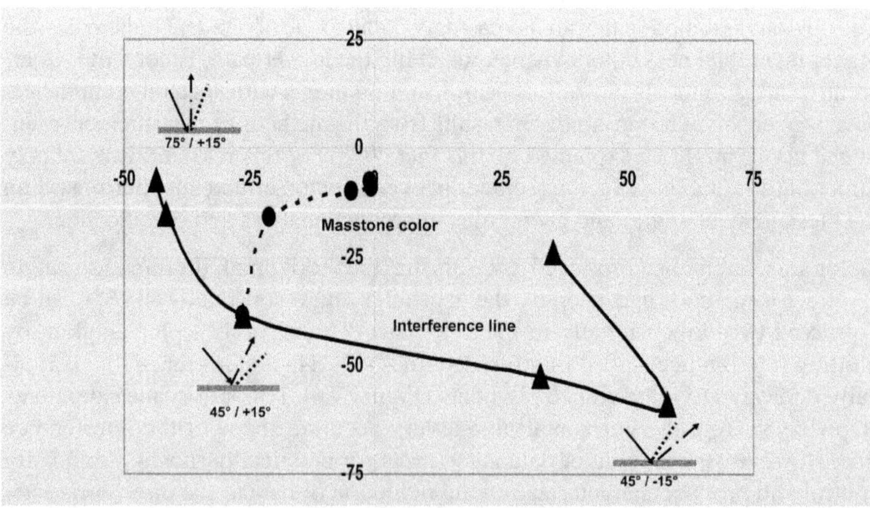

*Figure 7.7: Interference and aspecular shifts of a multilayer pigment based on mica in the a*b*-diagram measured by DATACOLOR-ETA system*

- For masstone the classical 45° illumination and the well-established 15°, 25°, 45°, 75° and 110° aspecular angles.
- For interference a set of three variable illumination angles 15°, 45° and 65° (related to the perpendicular!) as well as +15° and –15° aspecular observation angles.

7.4 Possible applications of colorimetric measurement for effect pigments

The most important applications of colorimetry with effect pigments are in process control and quality control of the manufactured pigments. The latter most take into consideration quality agreements between customer and supplier. In order to obtain meaningful results, i.e. metamerism it is necessary to operate with spectrophotometers, independent of the measurement geometry involved [15]. Two practical applications, which are at the beginning of their commercial development are discussed in the following. This includes the identification/characterization of transparent effect pigments in color formulations comprised of various types of pigment and the calculation of the resulting color shade based on colorimetric data.

The identification/characterization of effect pigments in color shades is not possible using colorimetric data alone. However, it does offer a useful and necessary supplement to analysis by microscope [20, 21].

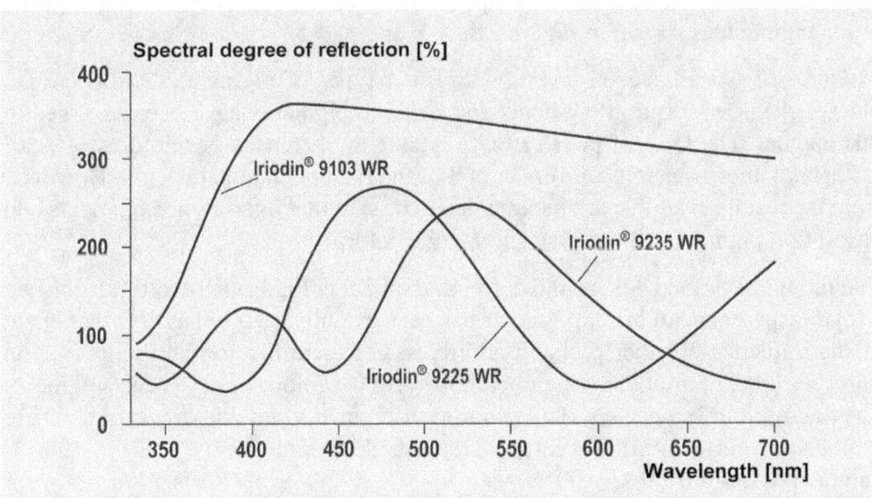

Figure 7.8: Spectral remission curves of a silverwhite titanium dioxide mica pigment and two interference pigments (blue and green)

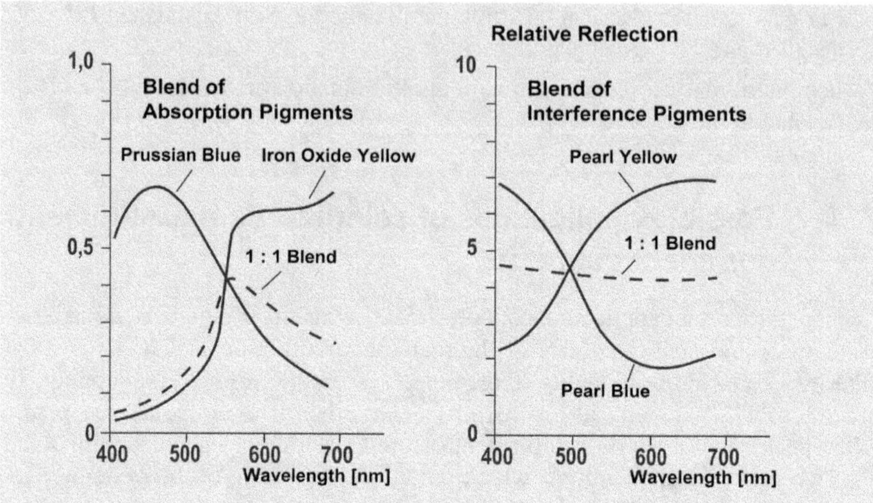

Figure 7.9: Laws of color mixing: subtractive for absorption pigments, additive for interference pigments

With the colorimetric data, more specifically with the color and light flop properties obtained using the two groups of measuring geometries; it is possible to clearly identify transparent effect pigments in a mixture of pigments. The reflection maximum/minimum shifts belong to different illumination angles which are typical for effect pigments. Furthermore, the shape of the curve can be seen as the "fingerprint" of the pigment, as shown in Figure 7.8.

Mixtures of pearl luster pigments with other types of pigment, such as carbon black, organic absorption pigments and metallic pigments have been assessed in this manner. The laws of physics that regulate the behavior of mixtures of pearl luster and other pigments are very complex and not understood in detail. However, from the results available at this time, it can be said that interference pigments and absorption pigments mix to give subtractive colors [22].

The influence of the background color on the final color shade of a film furbished with transparent pearl luster pigments has been examined as well as the assessment of the influence of other pigment additives. For identification and reproduction purposes it is important to remember that – also for colors – interference pigments obey additive mixing laws. The absorption pigments provide a contrast, where a blue and yellow mixture creates a green shade that means the color mixture is subtractive (Figure 7.9) [23].

For the identification of effect pigments in unknown paint formulation the measurement geometries of the new "FX10" are very useful [18]. Figure 7.10 points out

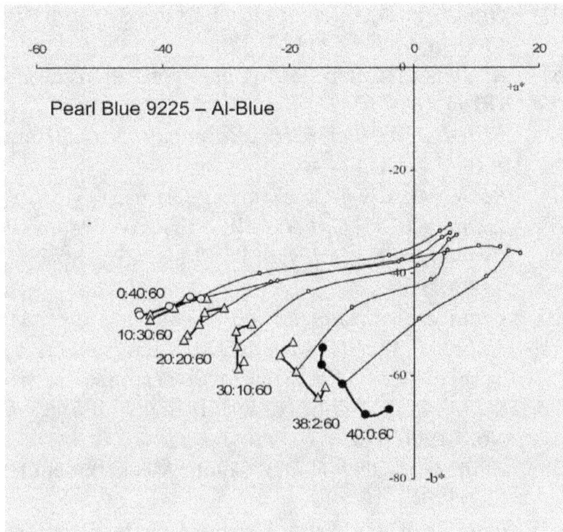

Figure 7.10: Aspecular lines from a mixture of interference-, aluminum- and absorption pigments show a parallel shift and follow the same direction. Aspecular lines are not suitable for identification of interference pigments. However, the interference lines are well observed with increasing concentration of the interference pigment in the mixture.

a*b*-results of a paint formulation with different concentration of blue absorption pigment, alumina pigment and an interference blue effect pigment. The measurement results show that the "aspecular line" from different aspecular angles measured by 45° illumination is not suitable to identify or to characterize the effect pigment in this paint formulation. Looking at the "interference line", measured by different illumination angles but constant +/– aspecular angle, the increasing concentration of the blue interference pigment is presented in the a*b*-diagram. Combining a*b*-values from the +/– aspecular angles a typical anchor shape is visible. The higher the concentration of the interference pigment the more marked the shape of the anchor.

This type of investigation has been repeated with various interference and absorption pigments. The results clearly demonstrate that interference effect pigments may be only well characterized by changing the illumination angle.

There is no fully developed commercially available system for the color calculation of interference color shades at this time. Several instrument manufacturers and pigment manufacturers are working to provide good, efficient and easy to usc systems to address this in the future.

References

[1] *Hofmeister, F.*, "Koloristische Bewertung von Perlglanzpigmenten", Farbe + Lack, **93** (1987) 799

[2] *Hofmeister, F., Pieper, H.*, "Einfluss einer variablen Meßgeometrie auf die Farborte von Interferenzpigmenten", Farbe + Lack, **96** (1990) 773

[3] *Gabel, P. W., Eberle, A., Parusel, M., Pieper, H.*, "Goniochromatic Quality Control of Effect Pigments", Conf. Proc., PRA, The Colour Delivery Challenge, Leeds, 2000

[4] *Gabel, P. W., Pieper, H.*, “Farbmessung von Interferenzpigmenten”, Technisches Messen, 60 Nr. 5 (1992) 226.
[5] *Rösler, G.*, “Colorimetric Characterization and Comparison of Metallic Paints”, PPCJ, **181** (1991) 230
[6] *Gabel, P. W., Pieper, H.*, “A comparison of different colorimeters”, Eur. Coat. J., 6/92, 356
[7] *Rösler, G.*, dfwg-Conference, Berlin 2005
[8] *Cramer, W. R., Gabel, P. W.*, “Effektvolles messen”, Farbe + Lack, **107** (2001) 42
[9] *Cramer, W. R., Gabel, P. W.*, “Measuring Special Effects”, Paint & Coating Industry, **9** (2001) 36
[10] *Cramer, W. R., Gabel, P. W.*, “Measuring Special Effects”, Eur. Coat. J., (7-8) (2001) 34
[11] *Gabel, P. W.*, “La Mesure Des Changements de Couleur Des Pigments A Effects Speciaux”, AFTPVA, Forum de la Connaissance, 2002, Paris
[12] *Gabel, P. W.*, “Time for a shift in colour assessment”, Asia Pacific Coat. J., Vol. 16, No. 1 (2003) 22
[13] *Pfaff, G., Gabel, P. W.*, “Colour Measurement from many angles”, Eur. Coat. J., (6) (2005) 30
[14] *Cramer, W. R.*, “Effekte sichten und beziffern”, Farbe + Lack, **108** (2002) 3
[15] *Gabel, P. W., Hofmeister, F., Pieper, H.*, “Interference Pigments as Focal Point of Colour Measurement”, Kontakte, Darmstadt, (2) (1992) 25
[16] *Hofmeister, F.*, “Reflectants Measurements on Absorption, Metallic and Mica Pigments”, Paint & Ink Intern., 4, Nr. 5 (1991) 17
[17] *Bäumer, W.*, “Perlglanz und Perlglanzpigmente”, Farbe + Lack, **79** (1973), 530, 574, 638
[18] *Cramer, W. R., Gabel, P. W.*, “Das gewisse Etwas, Dreiecksbeziehungen aus Bunt-, Aluminium- und Interferenzpigmenten”, Farbe + Lack, **10** (2004) 1
[19] DataColor, FX 10, company brochure
[20] *Hofmeister, F., Maisch, R., Gabel, P. W.*, “Farbmetrische Charakterisierung und Identifizierung von Mica-Lackierungen”, Farbe + Lack, **98** (1992) 503
[21] *Teaney, S., Denne, I.*, “Color Formulations with “Iriodin”, “Afflair”, Pigments for Automotive Coatings”, Kontakte, Darmstadt, (2) (1992) 46
[22] *Maisch, R., Gabel, P. W., Cramer, W. R.*, “New developments in the colour measurements of mica colours”, Proc. XIXth Intern. Conf. in Organic Coatings Science and Technology, Athen, 1993, 313
[23] *Hofmeister, F., Pieper, H.*, “Remissionsmessungen von Interferenz-, Aluminium- und Körperfarbenpigmenten”, Farbe + Lack, **95** (1989) 557

8 Impact of special effect pigments on humans and the environment

Dr. Gerhard Pfaff

8.1 Proof of physiological compatibility

In examinations for acute toxicity, no toxic effects were found for “Iriodin” pearl luster pigments (metal oxide mica pigments). Acute toxicity tests obtained showed the LD_{50} results to be greater than 15,000 mg/kg [1].

These metal oxide mica pigments demonstrate no irritation or sensitivity to contact with skin or mucous membranes [1–3]. Also for similar products based on other substrate materials such as silica flakes or borosilicate flakes, no irritation to the skin or mucous membranes was found.

Studies on rats regarding acute toxicity by inhalation of mica based effect pigments yielded LD_{50}-values ranging from 4.6 to 14.9 mg/l, larger 14.9 mg/l and larger 10.1 mg/l [4]. The animals probably died from over-exertion of the pulmonary systems (local effect) as a result of the ingestion of such a relatively large amount of material. A toxic effect due to the pigments could not be proven. Similar studies with pigments based on silica flakes also did not show any acute hazard through inhalation.

No hazard to humans has been found through contact with pearl luster pigments in the working place [5]. Studies on the genetic toxicity of pearl luster pigments have not shown mutagenic effects in vitro and in vivo. The results of Ames and microcore tests have not shown a mutation of the inheritance.

Chronic toxicity tests of mica based pearl luster pigments have resulted in absolutely no ill effects on rats over a 2 years period [6, 7].

Metal oxide mica pigments are highly stable and inert, and in aqueous suspensions virtually no metal ions can be leached from the surface. As a result this ensures that these pigments are non-toxic and do not cause any ecological hazard. As inorganic substances that are not soluble in water, pearl luster pigments are non-biodegradable.

8.2 Regulatory approval of metal oxide mica pigments

Cosmetics

Pigments especially recommended for use in cosmetic applications according to the current guidelines and regulations, for example “Timiron”, “Colorona”, etc. are registered for use in Europe (European Cosmetic Directive 76/768 EEC), the USA (21 CFR §73 and following), and Japan (New Japanese Cosmetic Regulation, April 1, 2001, PMSB Notification No. 990).

Food packaging/toys

All metal oxide mica based pigments without a special surface treatment are approved for use in food packaging, e.g. as a colorant in printing inks and plastics. Metal oxide mica based pigments fully meet the toxicity restrictions of European Resolution AP (89) 1 [8], which is the requirement for use of colorants in plastics in contact with foodstuffs. For use in Germany metal oxide mica based pigments without surface treatment fully comply with the recommendation IX of the federal institute for risk assessment (BfR recommendations, chapter IX [9]). Requirements for the same application in the USA is regulated according to CFR paragraphs 21 CFR §175.300 and following and §178.3297. In Japan, all pigments for use as colorants in the same application must pass a related approval procedure. The vast majority of "Iriodin" pigments are fully approved for food packaging in Europe and the USA and are also registered in Japan for this application. In addition, "Iriodin" pigments without surface treatment are approved under European Norm EN 71, part 3 for the use of colorants in toys [10].

References

[1] *von Eberstein, M., Rogula, P.*, "'Iriodin'-Prüfung auf akute Toxizität für Ratten nach peroraler Applikation und auf primäre Haut- und Schleimhautreizwirkung an Kaninchen", E. Merck, Darmstadt, 21. 05. 1970, Rep. No. 4/76/70

[2] *Heusener, A.*, "Test for skin sensitizing activity in the Open Epicutaneous Test (OET) in guinea pigs", E. Merck, Darmstadt, 08. 03. 1984, Rep. No. 4/15/84

[3] *Heusener, A.*, "Test for skin sensitizing activity in the Open Epicutaneous Test (OET) in guinea pigs", E. Merck, Darmstadt, 16.03.1984, Rep. No. 4/16/84

[4] *Therenaz, P.*, "4-hour, acute inhalation toxicity study with "Iriobronze" Aurum in rats", RCC, Itingen, Schweiz, Februar 1990, Rep. No. RCC Project 255767

[5] *Bruch, J.*, "Expert Report on Health Hazards caused by Pearl Lustre Pigments in Occupational Medicine and Toxicology", Institut für Hygiene und Arbeitsmedizin, Universitätsklinik Essen, 12.11.1990

[6] *Pence, D. H.*, "Combined oral chronic toxicity and oncogenicity study in rats with titanium dioxide coated mica", Hazelton Lab., America, Inc., Vienna, Virginia, 18.02.1987, Rep. No. 2164-100

[7] *Bernard, B. K., Osheroff, M. R., Hofmann, A., Mennear, J. H.*, "Toxicology and carcinogenesis studies of dietary titanium dioxide-coated mica in male and female Fischer 344 rats", J. Toxicology Environm. Health, **28** (1989) 415

[8] Council of Europe, Resolution AP (89) 1 on the use of plastic materials coming into contact with food, 13. September 1989 adopted by the committee of ministers at the 428th meeting of the ministers' deputies

[9] BfR-Recommendation (Bundesinstitut für Risikobewertung), Kapitel IX: "Farbmittel zum Einfärben von Kunststoffen und anderen Polymeren zum Einfärben von Bedarfsgegenständen" http://bfr.zadi.de/SEARCH/BASIS/kse1/all/blob_dt/DDD/090DEUTSCH.pdf.

[10] Directive 88/378/EWG from 3. Mai 1988 zur Angleichung der Rechtsvorschriften der Mitgliedsstaaten über die Sicherheit von Spielzeug; daraus abgeleitete Europa Norm 71, Teil 3: „Grenzwerte für die Migration bestimmter Elemente"

Acknowledgements

The first edition of this book, which first appeared in 1996 with the title “Special Effect Pigments”, would have never been published without the initiative of Dr. Roman Maisch. We would therefore like to express our thanks to him for paving the way for this present revised book.

We would particularly like to thank a number of Merck employees for their support with the successful preparation of this book: Dr. Michael Rösler for chapters 1 to 3, Dr. Peter Reynders and Dr. Kirsten Fritsche for chapter 3, Xenia Petsitis for chapter 6 and Ina Höfgen-Müller and Dr. Thomas Scheiring for chapter 8. The authors are particularly grateful to Monika Lenart for her highly faithful and discerning editing of the complete book and thanks also go to Helmut Plamper and Heinrich Opfermann for the preparation of the electron-microscopic images and to Dr. Ralf Glausch und Dr. Manfred Kieser, for their expert contribution to the chapter “Pearl luster pigments in paints” in the first edition of this book.

We would like to thank Dr. Josef Schmelzer (Kronos International, Inc., Leverkusen/Germany) and Dr. Jochen Winkler (Sachtleben Chemie GmbH, Duisburg/Germany) for making available the REM-images in figure 3.3 b/c and BASF AG in Münster/Germany as well as DuPont Performance Coatings GmbH in Wuppertal/Germany for figure 3.10. Thanks also go to Lutz Berthold and his colleagues from the Fraunhofer Institute for Mechanics of Materials in Halle/Germany for the preparation of the electron micrographs in the figures 2.15, 2.19, 2.58, 3.8, 3.9 and 2 (colored illustration).

Authors

Dr. Gerhard Pfaff studied chemistry at the Friedrich-Schiller-University in Jena, Germany. He specialized in inorganic, solid-state chemistry and obtained his PhD in 1983. Subsequently, he worked as scientific assistant and senior assistant in the faculty of chemistry of the same university, where he had a variety of roles in the teaching of inorganic chemistry. In 1991, Dr. Pfaff moved to Merck KGaA in Darmstadt, Germany. Since 1994, he was responsible for the product development of effect pigments. Since 2006 he is the global manager of pigments R&D. He gives lectures at the Technical University Darmstadt where he qualified as associate professor in the faculty of chemistry in 1997.

Dr. Peter Gabel studied physical chemistry at Technical University Braunschweig, Germany, where he also received his PhD. He joined Merck KGaA in 1986 and worked in the quality-control department in Darmstadt. In 1991, he moved to the Merck plant at Gernsheim, Germany where he became quality-control manager for pigments. For many years he has worked on the DIN color standards committee and the pigments and extenders standards committee, ASTM E12 color and appearance.

Dr. Frank J. Maile studied chemistry at the University of Stuttgart, Germany and gained his PhD at the Institute of Pigment and Paint Research (FPL) in Stuttgart and the Institute of Applied Macromolecular Chemistry of the University of Stuttgart in the field of applied electrochemistry using microprobing techniques. In 1998, he was awarded the Vincentz publishing house "Farbe und Lack" prize and in 2000 the paint and pigment section (APi) prize of the German Chemical Society (GDCh). From 1999 until 2005 he was responsible at Merck KGaA in Darmstadt for the servicing of car manufacturers in Europe and Latin America in pigment application engineering. Since 2006, Dr. Maile has been the group leader for technical service coatings in the PLS division of Merck KGaA.

Dr. Manfred Kieser studied chemistry at the University of Stuttgart, Germany and gained his PhD in 1971 in macro-molecular chemistry at the Institute of Pigment and Paint Research (FPL) in Stuttgart. Subsequently, he worked for another two years as a scientist at the institute. Since 1973, he has worked at Merck KGaA as laboratory manager in the field of effect pigments and their uses.

Joachim Weitzel began his career at Merck KGaA in 1969 with an apprenticeship in chemistry and physics. He went on to work on the research and development of the products and processes for cathodic sputtering for use in high-vacuum and ultra high-vacuum technologies. Since 1985, he ran the laboratory for grid electron microscopes for the analysis of pigments, photographic paints and for medical research projects. In 1990, he became the application engineering manager for pigments in printing applications and technical marketing manager of the printing group with the responsibility for international coordination. Since 1999, he is pigment application engineering manager at the Merck plant in Savannah in the USA.

Index

Symbole

A

B

C

D

E